DEVELOPMENTAL AND CELL BIOLOGY SERIES

EDITORS
M. ABERCROMBIE   D. R. NEWTH
J. G. TORREY

PRIMORDIAL GERM CELLS IN THE CHORDATES

PRIMORDIAL GERM CELLS
IN THE CHORDATES

Embryogenesis and phylogenesis

PIETER D. NIEUWKOOP

Hubrecht Laboratory, Utrecht, The Netherlands

LIEN A. SUTASURYA

Dept of Biology, ITB, Bandung, Indonesia

CAMBRIDGE UNIVERSITY PRESS

CAMBRIDGE

LONDON · NEW YORK · MELBOURNE

Published by the Syndics of the Cambridge University Press
The Pitt Building, Trumpington Street, Cambridge CB2 1RP
Bentley House, 200 Euston Road, London NW1 2DB
32 East 57th Street, New York, NY 10022, USA
296 Beaconsfield Parade, Middle Park, Melbourne 3206, Australia

First published 1979

Printed in Great Britain at the
University Press, Cambridge

Library of Congress Cataloguing in Publication Data
Nieuwkoop, Pieter Dirk
I. Sutasurya, Lien A., joint author. II. Title.
Primordial germ cells in the chordates
(Developmental and cell biology series)
Includes bibliographical references and index.
1. Germ cells. 2. Embryology – Chordata.
QL964.N53 1979   596'.03   78-18101
ISBN 0 521 22303 2

Dedicated with gratitude to
Professor Dr Chr. P. Raven, the former teacher of P.D.N.

Contents

viii   *Contents*

Preface

Since the extensive monograph of Bounoure (1939) on the origin of the primordial germ cells (PGCs) in the animal kingdom much work has been done, particularly on the ultrastructure of the PGCs in the insects and in the anurans. These studies provided strong support for the hypothesis of the existence of an uninterrupted 'germ line' throughout the entire animal life cycle, a concept first formulated by Weismann (1885, 1892) in his theory of animal heredity. In the two groups mentioned the germ cells are characterised by a particular cytoplasmic constituent, the so-called 'germinal plasm' or 'germ plasm', which is located in a particular region of the egg and can be traced through almost the entire life cycle of the individual. It is therefore not at all surprising that the concept of an uninterrupted germ line has become more and more generally accepted as a universal concept for animal development.

Ever since his first study on the origin of the PGCs in the urodeles in the nineteen-forties P.D.N. has had rather strong objections to a generalisation of the situation found in the insects and the anurans. The demonstration of the mesodermal origin of the PGCs in the urodeles and the first indications that the development of the PGCs could be due to inductive influences on the part of the ventro-caudal endoderm (Nieuwkoop, 1947) did not seem to be easily reconcilable with the situation in the anurans as described above.

The objections to this universal concept were markedly strengthened when Kotani (1957) suggested and Kocher-Becker & Tiedemann (1971) actually demonstrated that in the urodeles PGCs can be induced in blastula ectoderm by a heterogeneous mesodermal inductor. Subsequently it was shown that both in urodeles (Nieuwkoop, 1969a, b, 1970; Nieuwkoop & Ubbels, 1972; Boterenbrood & Nieuwkoop, 1973) and in anurans (Sudarwati & Nieuwkoop, 1971) the entire mesoderm develops epigenetically under an inductive influence emanating from the vegetal, endodermal moiety and spreading into the totipotent animal, ectodermal moiety of the blastula. The fact that in the urodeles the induced mesoderm may include PGCs (Boterenbrood & Nieuwkoop, 1973) gave these observations a much broader significance. This was further substantiated when L.A.S. actually demonstrated that in the urodeles the PGCs do not develop from predetermined 'elements', as in the anurans, but are a normal, characteristic constituent of ventro-caudal mesodermal differentiation and can be induced in any portion of the totipotent animal moiety of the blastula (Sutasurya & Nieuwkoop, 1974). It became evident that the situation encountered in the anurans is certainly not universal. At least two different principles are apparently applied in animal development to reach one and the same goal, that is, ensuring the survival of a species by means of sexual reproduction.

Although we know very little about the actual process of determination in germ cell development, it nevertheless seems obvious that germ cell development in the anurans is quite a different process from that in the urodeles. The fact that at least two different principles of germ cell formation exist among the vertebrates, and that they are encountered in the same taxonomic group, opens interesting new perspectives for introducing embryological evidence into the study of the phylogenetic relationships among the various groups of vertebrates. These new perspectives, both in the causal analysis of germ cell development and in the study of phylogenetic relationships among the vertebrates, constitute the motive for writing this monograph. The authors hope that its publication will actually stimulate new research in this fascinating field, particularly in the hitherto neglected groups.

Acknowledgements

First of all we want to thank the editors of this series of Cambridge University Press for the invitation to write this monograph.

We are very grateful to the Department of Biology of the Institut Teknologi Bandung for granting leave to L.A.S. to return to Holland for literature study for the book as well as for the preparation of the final manuscript.

We also extend our thanks to the Governing Board of the Hubrecht Foundation for repeatedly providing the necessary funds for the travel of L.A.S.

We want to thank the staff of the library of the Hubrecht Laboratory for their assistance in compiling the extensive bibliography, Mrs E. Wolters-Bartová for preparing most of the illustrations, and Miss S. J. de Vos for typing the manuscript.

We express our deep appreciation to Dr J. Faber for his valuable suggestions and for the correction of the English text.

Finally we are indebted to several colleagues for allowing us to reproduce figures or for sending us original illustrative material.

1

General introduction

Statement of problems

Cambar (1956) characterised development as the phenomenon that the spatial multiplicity of a system gradually increases by processes of cytoplasmic segregation and cellular interaction. The latter are markedly intensified by extensive morphogenetic movements which bring into contact regions of the embryo which originally were far apart.

The result of development is a highly complex new individual which is a copy of the parental organisms. Like its parents it consists of two main components, viz. the somatic cells of many different types, and the germ cells. The somatic cells constitute the body of the individual of a particular generation, whereas the germ cells represent the forerunners of the next generation. These theoretical considerations led to the fundamental distinction between '*soma*' and '*germen*'.

The extensive early literature on the origin of the germ cells reflects in part advocacy and in part rejection of this concept. Waldeyer (1870) did not ascribe any importance to the distinction between *soma* and *germen* and had the germ cells originating from somatic cells of the 'germinal epithelium' of the gonadal anlage. Nussbaum (1880) was the first to advocate an early segregation of the germ cells from the somatic cells of the organism. Weismann gave new impetus to the distinction between *soma* and *germen* in his theory of the functioning of heredity in development, laid down in his two classical works of genius: *Die Continuität des Keimplasmas als Grundlage einer Theorie der Vererbung* (1885) and *Das Keimplasma. Eine Theorie der Vererbung* (1892). He did not, however, identify the continuity of the germ plasm with an early segregation of the germ cells – he studied, among other things, the late appearance of the germ cells in the hydroid medusae (1883). However, the notion of an early segregation of the germ cells from the somatic cells fell in line so well with his theory that an association of the two ideas was unavoidable. This led to a more isolated position for Waldeyer's concept, since an origin of the germ cells from differentiated somatic cells of the 'germinal epithelium' of the gonadal anlage is not in accordance with Weismann's theory.

Bounoure (1939), in his book *L'origine des cellules reproductrices et le problème de la lignée germinale*, reviewed the descriptive and experimental evidence for the two opposing points of view in the various groups of the animal kingdom from the time of Waldeyer till the late nineteen-thirties. On the basis of his own research on the origin of the germ cells in the anuran amphibians he advocated the very early segregation of the germ cells from the somatic cells, as well as their continuity throughout the life cycle.

Cambar (1956) formulated the two opposing points of view as follows:

(*a*) The embryo initially consists of somatic cells only; at this stage the germ cells are neither differentiated nor determined. The future germ cells therefore share the common fate of the somatic cells during a long period of development. They finally segregate from the somatic cells by a late, regional chemo-differentiation under the influence of external inductive actions. *Germ cell differentiation is a typically epigenetic phenomenon.*

(*b*) The germ cells segregate from the somatic cells during the first cleavages of the egg and represent a true 'germ line', which is distinct from the lineage of the somatic tissues from the very beginning of development. The region of the egg from which the germ cells originate is characterised by the presence of a special cytoplasmic structure, the so-called 'germ plasm', which acts as a germinal determinant and ends up exclusively in the germ cells as a result of differential cell divisions. Since the early segregated germ cells constitute the only source of the future gametes, there exists a germinal continuity through successive generations. *The germen is unique and irreplaceable; it represents a typically preformistic element in development.*

The first alternative holds, for example, for the hydroid medusae, where sexual and asexual generations alternate, as well as for the echinoderms, where gonad and germ cells appear only late in development. The second alternative holds, for example, for the anuran amphibians and for several insects, where segregation of the germ cells actually occurs during cleavage and where an uninterrupted germ line can be demonstrated.

We know, however, that in addition to these typical cases a great variety of other situations is encountered among the invertebrates and vertebrates. Focussing our attention on the chordates, a typical germ plasm has hitherto not been found in the higher vertebrates although germ cells can be identified at early stages of development. The same holds for the urodeles, where it has been demonstrated that the germ cells can arise from common, totipotent cells of the animal ectodermal moiety of the blastula and early gastrula (see chapter 5, p. 91).

We must therefore ask ourselves whether we have not fallen victim to our own conceptual thinking. Must we not reconsider the question of whether the two opposing views are actually as mutually exclusive as they seem at first sight, and whether they are theoretically the only possible ones.

First of all it should be realised that embryonic cells which contain germinal plasm cannot be considered to be true germ cells so long as they still contain a certain amount of 'non-germinal' or somatic cytoplasm. Only those cells which have fully segregated from the future somatic cells will actually give rise to germ cells when placed in the proper environment. This, however, does not necessarily mean that at the end of the segregation period these cells are already irreversibly determined to become germ cells. We know hardly anything about the time and nature of germ cell determination. In molecular terms germ cell determination is not definitive until the specific transcription for germ cell differentiation begins.

Within the framework of the concept of the continuity of the germ line and the segregation of *germen* and *soma* the distinction between cells with germinal determinants and true germ cells may seem academic. However, this distinction acquires a broader significance when the concept of true somatic cells is also considered. For one can only speak of true somatic cells when the specific transcription for a certain type of somatic differentiation has begun. The initial phase of development, before somatic and germ cell differentiation has started, must therefore be considered as a separate period during which no part of the original zygote is yet determined for a particular pathway of differentiation. This phase may be called *the totipotent phase of development*. In some groups, such as the anurans, this may be restricted to the very first cleavage divisions of the egg, while in others it may last much longer. It may, for example, still apply to the single-layered chick blastoderm (see chapter 2, p. 42). Totipotency may moreover be restricted to a particular portion of the embryo, for instance the animal ectodermal moiety of the urodelan blastula and early gastrula, from which in addition to all somatic cell types germ cells may also develop (see chapter 2, p. 12 and chapter 5, p. 90). The same totipotency, however, does not seem to hold for the animal ectodermal moiety of the anuran blastula or early gastrula, from which as far as we know all somatic cell types but no germ cells can be formed. Following this line of thought it must be emphasised that so long as there are totipotent cells in an embryo, or in an adult organism, one of the primary requirements for germ cell formation seems to be fulfilled. It is interesting to note that these considerations fall entirely within the framework of Weismann's *Keimplasma* theory. Only in a small number of invertebrates – i.e. in the Ascaridae, Diptera and Hymenoptera, and Crustacea – does chromosome elimination or chromatin diminution occur. In the great majority of animal groups, and surely in all chordates, the full complement of genetic information present in the zygote is transferred to all cells of the embryo (see Gurdon, 1974). In these groups cellular differentiation must represent a process of selective repression or derepression of a particular segment of the genome. Since there are indications that this process is not necessarily irreversible, dedifferentiation may actually lead to totipotency again. This would mean that germ cell formation during late development and even in the adult seems at least feasible without the necessity of totipotent cells being preserved during the entire period between fertilisation and actual germ cell formation.

Having arrived at this conclusion a number of intriguing questions arise. In the anuran embryo the germ cells originate from the vegetal pole region of the egg, where the germ plasm is located. Experimental evidence strongly suggests that the yolk-rich vegetal region of the egg is precociously determined to form part of the endoderm of the future embryo (see chapter 2, p. 12). It is now generally accepted that in the anurans the germ cells are of endodermal origin. Apart from the presence of the germ plasm the presumptive germ cells are indistinguishable from the surrounding endodermal cells. Do the presumptive germ cells in the anurans actually go through a phase of somatic (endodermal) differentiation – however short it may be – before being switched into the ultimate pathway of germ cell differentiation? If so, what does the purported totipotency of the presumptive germ cells actually mean? If not, could the role of the germ plasm merely be to preclude a temporary phase of somatic determination or, in other words, does the germ plasm prevent the loss of totipotency of the presumptive germ cells in an environment which favours somatic differentiation?

Considering the situation in the urodeles the following questions arise. Are the presumptive germ cells in the urodeles characterised, as in the anurans, by the presence of germinal plasm prior to their determination? If so, when and how does the germinal plasm arise in development? If not, how does the determination of the presumptive germ cells occur in the absence of germinal plasm? Which is more essential for germ cell development, the maintenance or re-establishment of totipotency or the presence of a germinal determinant?

It is generally accepted that both spermatozoa and egg cells represent highly specialised cell types. The spermatozoon has a very complex cytoplasmic structure upon which its specific functions of chemotactic motility and ability to penetrate a mature egg cell are based. Apart from its receptivity to the male gamete, the egg cell must be fully prepared for autonomous development in an essentially hostile environment. This requires specific properties of its cell membrane and cytoplasmic machinery. It must moreover be able to undergo the very complex processes of spatial differentiation and organisation that we call development. When totipotency turns out to be the principal characteristic of primordial germ cells, how must germ cell differentiation be envisaged? Can a cell be highly differentiated without losing its totipotency?

Focussing our attention upon the apparent discrepancy in germ cell formation between the anurans and the urodeles, other questions present themselves. Are there only two different modes of germ cell formation, at least among the chordates, or can still other modes be distinguished? If so, what have these various modes of germ cell formation in common? If not, which of the two modes occurs in the different groups of fishes and reptiles and which in birds and mammals? Do the differences in germ cell formation among the various groups actually reflect differences in their phylogenetic origins? What can embryology in general contribute to our present insight into the phylogeny of the chordates?

In the chapters 3 to 7 we hope to approach some of the questions concerning the origin, structure and functional significance of the characteristics of the germ cells by reviewing the literature relating to the various groups of chordates over the last four decades. For the older literature one may refer to the excellent review by Bounoure (1939). For a better understanding of the process of germ cell formation in the various groups we intend to place it against the background of our present insight into early development (chapter 2). The possible phylogenetic significance of these data will be discussed in chapter 8. It must be realised, however, that in several groups relevant data are still very scanty or even completely lacking. Finally, in chapter 9 suggestions will be made for appropriate further analysis.

It was originally intended to restrict the monograph to the vertebrates, but some very interesting parallels between the early development of amphioxus and that of the amphibians have led us to extend the study to the entire phylum of the chordates.

Although Bounoure's book covers the entire animal kingdom we considered it more appropriate to restrict the present book to the chordates. Inclusion of the invertebrates would unavoidably lead to a considerable increase in size, since about half of Bounoure's book is devoted to the invertebrates. Such expansion would make the book less readable. Moreover, phylogenetic considerations, which seem acceptable for the rather coherent phylum of the chordates, would become completely speculative and unjustified when extended to the entire animal kingdom.

Since it was our intention to focus attention upon some of the fundamental problems concerning the origin and determination of the germ cells, we decided to restrict ourselves to the early history of the germ cells, from their first origin to their indifferent state in the sexually undifferentiated gonad (see table 1.1). This means that the book will only deal with primordial germ cells, leaving their further history in the sexually differentiated gonad out of account. Their later history is so many-sided that oogenesis and spermatogenesis ought to be discussed separately for an adequate treatment.

Terminology of gametogenesis and gonad formation

The non-specific term 'germ cells' will be used indiscriminately for all stages in the formation of the germ cells.

In the chordates germ cells are encountered (1) during the greater part of embryonic development, (2) during the entire larval period (when such a phase of development can be distinguished), and (3) during the entire or almost the entire period of adult life. The uninterrupted presence of germ cells during almost the entire life span of the individual has led to the concept of the 'germ line' or '*germen*' in contradistinction to that of the 'somatic line' or '*soma*'.

The development of the germ cells is evidently a long and complex process, in which various phases can be distinguished. Unfortunately, different authors have not always used the same terminology. It therefore seems desirable to define the various phases, particularly those which will be dealt with in this book.

The entire development of the germ cells, from their first detectable origin until their release as mature gametes, is called 'gametogenesis'. This long period is usually subdivided into two major periods, that before and that after sexual differentiation. During the former period, i.e. before the sex of the individual becomes phenotypically distinguishable in either the gonadal anlage or the germ cells themselves, the cells will be called 'primordial germ cells' (PGCs), which is synonymous with the German term '*Urgeschlechtszellen*' and the French term '*gonocytes*'.

Within this period we must specifically distinguish the phase between the first identification of the germ cells and the beginning of their morphological differentiation. In groups such as the anurans this phase comprises the segregation of the germ cells and the process of determination. In other groups, such as the urodeles, where segregation of the germ cells from the somatic cells cannot be demonstrated, it comprises at least the process of determination prior to cytoplasmic differentiation. It therefore seems advisable to designate the forerunners of the PGCs in this phase as 'presumptive primordial germ cells' (pPGCs), in contradistinction to the subsequent phase of 'true primordial germ cells' (PGCs) (table 1.1, right-hand column).

Several types of PGC, such as those of the anurans and certain insects, are characterised by the presence of the so-called 'germinal plasm', a particular cytoplasmic constituent which is considered to be specific for germ cell development and is assumed to act as a 'germinal determinant' (Hegner, 1911). Synonyms are 'germ plasm' and 'germinal cytoplasm'. In other types of PGC a so-called 'nuage material' with rather similar ultrastructure is found.

As soon as sexual differentiation sets in, male and female individuals can be distinguished. The entire process of egg cell formation, starting with the initiation of the first meiotic division, is called 'oogenesis'. The germ cells in the female gonad are initially called 'oogonia'. When meiosis starts the female germ cells enter the 'oocyte' stage, which comprises the cytoplasmic growth phase, the entire period of vitellogenesis, and the ultimate maturation of the egg cell (Henderson & Henderson, 1953; Baker, 1972). The mature female gamete or 'ovum' is also called the 'mature, unfertilised egg'. Similarly, in the male gonad the entire process of germ cell formation is called 'sperma-

Table 1.1. *Phases in the development of the germ cells, with special reference to their designations in different periods of gametogenesis (see text for further explanation)*

I. Early development up to sexual differentiation	
(a) Extragonadal period	
1. Segregation and determination	pPGCs
2. Morphological differentiation	PGCs
3. Migration	PGCs
(b) Formation of gonadal anlage	
4. Colonisation	PGCs
(c) Indifferent gonad	
5. Multiplication	PGCs
II. Sexual differentiation	
(d) Differentiation in ovary and testis	Oogonia, spermatogonia
6. Oogenesis and spermatogenesis: meiosis	Oocytes, spermatocytes
III. Sexual maturation	Ova, spermatozoa
IV. Fertilisation	Zygote

togenesis'. The germ cells are initially called 'spermatogonia'. When meiosis starts they are named 'spermatocytes', while the mature male gamete is called 'spermatozoon'.

After the union of the ovum and the spermatozoon the second meiotic division of the ovum is completed and the second polar body extruded. When the haploid male and female pronuclei fuse the 'zygote' stage is reached and development of the next generation starts (table 1.1, right-hand column).

In the vertebrates the 'gonadal anlagen' first appear as longitudinal folds of the coelomic epithelium on either side of the dorsal mesentery. This epithelium is currently called 'germinal epithelium', but this term should be replaced by the more accurate term 'gonadal epithelium' (see chapter 6, p. 104). These folds soon take the form of longitudinal ridges protruding into the coelomic cavity and are called the 'genital ridges' or 'gonadal ridges'. They are found in a caudal portion of the trunk, in the region of the mesonephric anlagen. After colonisation with PGCs the gonadal anlage develops into the 'indifferent gonad' of the sexually undifferentiated embryo, which shows a bisexual character with 'cortical' and 'medullary' components. Sexual differentiation comes about either by the predominant development of one of the two components in either sex, or by a preceding displacement of the germ cells from the cortical into the medullary component in the male. The gonadal anlage of the female is called the 'ovary' and that of the male the 'testis' (table 1.1, left-hand column).

2

Early embryogenesis, with special reference to mesoderm formation

Embryogenesis, and in particular mesoderm formation, is not equally well understood in the various groups of the chordates. Our insight actually varies tremendously; some groups are reasonably well studied, while our knowledge of others is very scanty. Therefore, a treatment of the subject strictly according to the present taxonomic classification of the chordates (table 2.1) seems inappropriate. It is more logical to start with a description of the well-known groups and to proceed from there via the less thoroughly studied to the least-known groups.

The phylum Chordata is subdivided into three subphyla, the Tunicata, the Cephalochordata, and the Vertebrata; the Vertebrata is again subdivided into eight classes, of which the first five comprise the Anamnia and the remaining three the Amniota (table 2.1).

The group whose embryogenesis is best understood and in which mesoderm formation has been studied most thoroughly is doubtless the urodele amphibians, although quite a lot is also known about the anuran amphibians. Among the lower chordates, the Cephalochordata show very interesting parallels with amphibian development, whereas little is known about mesoderm formation in the Tunicata. Among the fishes the Agnatha and Osteichthyes have an early development which is rather similar to that of the amphibians. Less well studied, particularly experimentally, are the Chondrichthyes and the rather aberrant Teleostomi. Among the amniotes avian development is by far the best understood, the eggs being reasonably accessible to experimental intervention. The recent development of in-vitro techniques for early mammalian embryos has made mammalian development a promising new field of research. Our present insight is still rather scanty, however. The reptiles are the most neglected group among the amniotes, for which hardly any experimental data exist.

All vertebrate groups have in common that the PGCs migrate from their extra-gonadal site of origin towards the mesodermal gonadal anlagen, passing through various endodermal or mesodermal tissues (see chapter 7, p. 113). In the urodele amphibians the PGCs originate in the presumptive lateral plate mesoderm and their formation seems to be closely associated with the formation of the mesoderm itself (see chapter 5, p. 88).

Table 2.1. *Taxonomy of the chordates*

Phylum Chordata
    Subphylum Tunicata
        Class Ascidia   Order Enterogona
    Subphylum Cephalochordata
        Class Leptocardii   Order Amphioxiformes
    Subphylum Vertebrata
      The Anamnia
        Class I Agnatha   Order Petromyzoniformes
        Class II Chondrichthyes
            Order I Squaliformes or Selachii
            Order II Rajiformes or Batoidei
        Class III Osteichthyes
            Order I Dipteriformes or Dipnoi
            Order II Polypteriformes
            Order III Acipenseriformes
            Order IV Lepisosteiformes
            Order V Amiiformes
        Class IV Teleostomi
            Orders I–XXX (Greenwood *et al.*, 1966)
        Class V Amphibia
            Order I Urodela or Caudata
            Order II Coecilia or Apoda
            Order III Anura or Salientia
      The Amniota
        Class VI Reptilia
            Order I Chelonia
            Order II Crocodilia
            Order III Squamata
        Class VII Aves
            Orders I–XXVI (Grzimek, vol. VII–IX, 1968–70)
        Class VIII Mammalia
            Orders I–XVIII   (Grzimek, vol. X–XIII, 1968–72)

The Anamnia

Amphibia

Among the meso-lecithal amphibian eggs, different types can be distinguished according to the yolk content. Among the Anura the variation in yolk content is relatively small. The eggs of species which have a short embryonic development, such as *Xenopus* species, are relatively poor in yolk. Other eggs, like those of *Rhacophorus* species, are much richer in yolk but still show a holoblastic cleavage type, albeit with a rather long lag in the fragmentation of the vegetal half. Among the Urodela a much greater variation in yolk content is found. The eggs of *Triturus* or *Ambystoma* species contain relatively little yolk, while those of *Necturus* and *Cryptobranchus*, and particularly those of *Salamandra*, are very rich in yolk. The latter approach the meroblastic cleavage type, where cleavage initially occurs only in the upper, animal region, the fragmentation of the vegetal moiety being markedly retarded. The coecilian egg is very rich in yolk, so that cleavage is initially of the meroblastic type, leading to the formation of a blastodisc which covers about one-fifth of the egg surface but is not sharply delimited. The yolk mass, however, finally becomes fragmented into a number of large blastomeres, so that cleavage is not truly meroblastic.

Urodela

The urodele egg shows a pigmented animal half and an unpigmented vegetal half. Shifts in the pigment distribution of the fertilised egg shortly before the onset of cleavage lead to the formation of a so-called 'light crescent' which constitutes the first externally visible sign of the establishment of dorso-ventral polarity. The process of symmetrisation is less well understood in the polyspermic urodele egg than in the monospermic anuran egg, and what determines the plane of bilateral symmetry is essentially unknown. We only know from local insemination experiments that, as in the anurans, the dorsal side usually appears opposite the place of sperm entry (Ancel & Vintemberger, 1948). The role of the gravitational rotation of the freshly laid egg has been thoroughly studied only in the anuran egg. The influence of gravity may also play a role in the presymmetrisation of the egg during oogenesis.

The urodele egg shows holoblastic cleavage, becoming subdivided into large blastomeres. The first two cleavages are meridional, dividing the embryo into four nearly equal blastomeres. The third cleavage is equatorial, the animal blastomeres being markedly smaller than the vegetal ones. Since in every cleavage cycle all the blastomeres divide, with the vegetal blastomeres lagging only slightly behind the animal ones – a so-called synchronous cleavage pattern – the number of cells increases exponentially and reaches about 2000 after the eleventh cleavage (Hara, 1977). Then synchrony is gradually lost as a result of a lengthening of the cleavage cycle of individual blastomeres.

From the third cleavage onwards a blastocoelic cavity is formed by the excretion of fluid into the interblastomeric spaces. It increases gradually in size, reaching its maximal extent in the late blastula and early gastrula. The blastula is essentially a hollow, single-layered sphere, all cleavage planes being oriented more or less perpendicular to its surface.

The blastocoelic cavity divides the embryo into two different moieties, a massive vegetal, yolk-rich moiety and a relatively thin-walled, cap-shaped

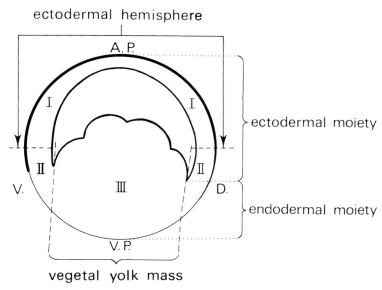

ectodermal hemisphere

Fig. 2.1. Diagram of the subdivision of the amphibian blastula into two functionally different moieties and three successive regions, viz. animal, ectodermal hemisphere (I), subequatorial zone (II) and vegetal yolk mass (III). A.P., animal pole; D., dorsal; V., ventral; V.P., vegetal pole.

animal moeity (see fig. 2.1). Experiments involving isolation of the two moieties at a late morula/early blastula stage lead to a different development of the two parts. The vegetal moiety forms a yolk-rich, undifferentiated atypical 'endodermal' mass which stops developing after a few further divisions, while the animal moiety gives rise to an irregular, atypical 'ectodermal' cell mass which forms cilia but then stops developing (Grunz, Multier-Lajous, Herbst & Arkenberg, 1975). Isolation of the two moieties at a later blastula stage leads to mesoderm formation in the animal ectodermal moiety (Nieuwkoop, 1969a). Subdivision of the late blastula into an animal hemisphere (I), a subequatorial ring-shaped portion (II) and a massive vegetal yolk mass (III) results in development of (I) and (III) similar to that of the two moieties of the late morula/early blastula, while the subequatorial portion (II) gives rise to the endodermal and mesodermal and sometimes also some ecto-neurodermal structures. Recombination of the animal hemisphere with the vegetal yolk mass leads again to mesoderm formation and to the development of all endodermal, mesodermal and ectoneurodermal structures. In some cases completely normal embryos arise from such recombinates (Nieuwkoop, 1969a).

Using xenoplastic recombinates of *Triturus* and *Ambystoma* blastulae or recombinates of labelled and unlabelled eggs of *Ambystoma*, Nieuwkoop &

Ubbels (1972) demonstrated that in these recombinates all the mesodermal structures were formed from the animal hemisphere under an inductive influence emanating from the vegetal yolk mass, which itself formed only endodermal structures. In cases where the large circular wound of the recombinate healed completely before the onset of gastrulation and as a consequence gastrulation proceeded more or less normally, pharyngeal endoderm was formed from the animal hemisphere in addition to all the mesodermal structures.

When recombinates of animal hemispheres and vegetal yolk masses were treated with lithium chloride as a 'vegetalising agent', the competence of the animal hemisphere for mesoderm and endoderm formation was so strongly enhanced that nearly the entire animal hemisphere formed endoderm and mesoderm, leaving only a small fraction for ectodermal and neurodermal development (Nieuwkoop, 1970). Similar results had previously been obtained by Kocher-Becker, Tiedemann & Tiedemann (1965) with early gastrulae of *Triturus*, after injection into the blastocoelic cavity of a purified 'mesodermal inductor' isolated from chick embryos. In contrast, when so-called 'animalising agents' such as zinc chloride and mono-iodo-acetate are used, cells of the vegetal yolk mass are never transformed into mesodermal or ecto-neurodermal structures, showing that in the urodele blastula the vegetal yolk mass is already firmly determined for an endodermal pathway (Nieuwkoop, 1973).

Although the possibility of the existence in the embryo of a small transitional region between the animal and vegetal moieties (as suggested by Koebke, 1974) cannot be excluded, the experiments cited above strongly suggest that in the urodeles mesoderm formation is a typically epigenetic phenomenon and that the blastula consists essentially of two components: a totipotent animal moiety and a vegetal moiety whose developmental potencies are already restricted to endodermal development (see Grunz, 1975).

Studying the differentiation tendencies of the presumptive mesoderm in small fragments of the dorsal, lateral or ventral marginal zone at successive stages, Nakamura & Matsuzawa (1967) found that mesodermal differentiation tendencies appear in the marginal zone at a late morula/early blastula stage. Although mesodermal structures may be induced in gastrula ectoderm by heterogeneous inductors (Toivonen, 1953) and likewise by the vegetal yolk mass (Nieuwkoop, 1969a), Nakamura & Takasaki (1971) nevertheless believe that in normal development mesoderm formation is due to a spatial interaction between an animal and a vegetal gradient as demonstrated by Runnström (1967) in the sea urchin. However, the fact that, at least at the blastula stage, the embryo cannot be animalised but only vegetalised, pleads against such a double gradient hypothesis and is a strong argument for the thesis that the mesoderm is actually induced in the animal moiety by the vegetal endodermal yolk mass (Nieuwkoop, 1973).

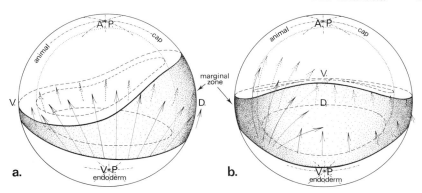

Fig. 2.2. Formation of ring-shaped mesendodermal marginal zone in the amphibian blastula as a result of inductive action emanating from the peripheral region of the vegetal endodermal moiety and spreading with decrement into the animal ectodermal cap. There is strong dorsal and weak lateral and ventral inductive action. Three-dimensional projection: (a) lateral view, (b) dorsal view. A.P., animal pole; D., dorsal; V., ventral; V.P., vegetal pole. (See Weyer *et al.*, 1977.)

Translocation experiments at the blastula stage in which the animal hemisphere was rotated through 90° or 180° with respect to the vegetal yolk mass (Nieuwkoop, 1969*b*), demonstrated that at this stage the dorso-ventral polarity of the embryo resides in the vegetal yolk mass and not in the animal ectodermal hemisphere, notwithstanding the fact that the latter contains a part of the light crescent. Moreover, recombinates of separate dorsal, lateral and ventral portions of the yolk mass with animal hemispheres from early to middle blastulae showed that the dorsal region of the yolk mass contains a strong, dominant centre for mesoderm induction, whereas in both the lateral and the ventral region this capacity is only weak (Boterenbrood & Nieuwkoop, 1973). The dorsal centre predominantly induces axial mesoderm, particularly notochord and somites, while the lateral and ventral portions of the yolk mass induce ventral mesodermal structures, such as blood islands and lateral plate mesoderm. The regional differences in inductive capacity of the yolk mass are reflected in the varying width of the marginal zone around the circumference of the egg: this zone, which represents the presumptive mesodermal mantle, is markedly broader dorsally than ventrally, with an intermediate width laterally (fig. 2.2). It must, however, be emphasised that some ventral mesodermal structures may also be formed in the 'dorsal' recombinates, and that pronephric tubules, belonging to the intermediate mesoderm, are formed in the 'dorsal', but particularly in the 'lateral' and 'ventral' recombinates. These data suggest that the differences in mesoderm-inducing capacity found within the yolk mass are initially essentially of quantitative character. In subsequent development the inducing action may, however, gradually acquire some qualitative regional aspects.

Mesodermal differentiation tendencies thus appear very early in development and result from an interaction of the two different moieties of the embryo. This requires both an early appearance of inducing capacity in the yolk mass and an early development of mesodermal competence in the animal hemisphere. It is known from Leikola's (1963) experiments with a heterogeneous inductor (guinea pig bone marrow) that the animal moiety of the *Triturus* embryo loses its mesodermal competence at the early to middle gastrula stage (stage 11, Harrison, 1969). Boterenbrood & Nieuwkoop (1973) found in *Ambystoma* that the mesoderm-inducing capacity disappears from the vegetal yolk mass at an earlier stage. In its dorsal portion the inducing capacity begins to decline at the late blastula stage (stage 9) and has disappeared completely with the first appearance of the dorsal blastoporal groove (stage 10^-). At that stage the lateral portions of the yolk mass only have a reduced inductive capacity, while that of the ventral portion is still almost unaffected. It is therefore evident that mesoderm induction is a process of rather long duration.

The long duration of mesoderm formation becomes even more evident when the mesodermal differentiation tendencies within the marginal zone at the end of the initial induction period are analysed. Holtfreter-Ban (1965) demonstrated that at the early gastrula stage the dorsal marginal zone still does not show a well-defined regional pattern of mesodermal differentiation tendencies. Almost any region can still form a great variety of both mesodermal, and ectodermal and neurodermal structures. Its 'caudal' half shows markedly weaker differentiation tendencies than its 'cranial' half, which borders on the dorsal blastoporal groove. In *Cynops pyrrhogaster* Hama (personal communication) failed to find any mesodermal differentiation tendencies in the 'caudal' half of the dorsal marginal zone at the early gastrula stage. It is therefore evident that mesoderm formation is far from completed in the early gastrula and that it requires further inductive interactions before the final pattern of the mesoderm is formed.

Before discussing these interactions the complex changes of form that take place during the gastrulation process must be described in bird's-eye view. By means of coordinated cell movements both the marginal zone material and the vegetal yolk mass are carried from the outer surface of the egg into its interior, transforming the initially single-layered blastula into a triple-layered embryo. This invagination process is preceded by a gradual extension of the animal moiety in the vegetal direction ('epiboly') and by a movement of individual blastomeres from the vegetal surface into the interior of the egg ('local ingression'). These two phenomena constitute the 'pregastrulation' movements (Schechtmann, 1934). The gastrulation process itself was extensively described by Vogt (1929). His original anlage maps of the blastula and early gastrula have been partially supplemented and corrected by Nakamura (1942) and Pasteels (1942) (cf. the urodele anlage map, fig. 2.3).

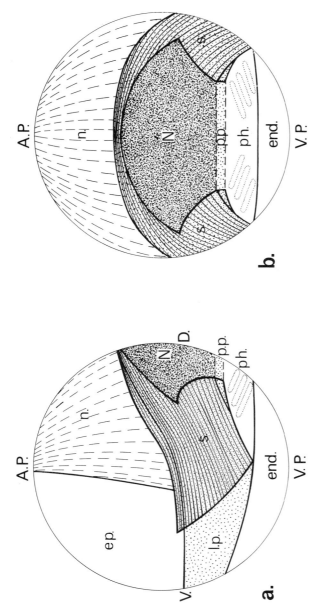

Fig. 2.3. Organ anlage map of the urodele gastrula (external marginal zone). (a) Lateral view, (b) dorsal view. A.P., animal pole; D., dorsal; end., endoderm; ep., epidermis; l.p., lateral plate; n., neural plate; N, notochord; ph., pharynx endoderm; p.p., prechordal plate; s., somites; V., ventral; V.P., vegetal pole. (After Vogt, 1929, as corrected by Nakamura, 1942, and Pasteels, 1942.)

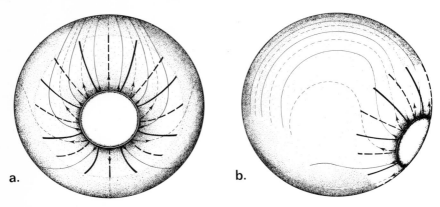

Fig. 2.4. Diagram of movements of the marginal zone during the formation of archenteron roof (external displacements: heavy lines; internal displacements: thin lines), showing dorsal convergence and ventral divergence as well as longitudinal stretching. (After Vogt, 1929.)

The process of invagination starts on the dorsal side, at the boundary of the vegetal yolk mass and the light crescent, giving rise to a slit-shaped deep invagination, the future archenteron. Subsequently the invagination spreads from the dorsal towards the lateral and ventral sides, so that a circular groove is finally formed around the vegetal yolk mass, the 'yolk plug'. The yolk plug is slowly pulled inwards together with the advancing, now internally situated mesodermal mantle. Directed cell movements with lobopodium formation occur at the advancing edge of the endo-mesoderm (Nakatsuji, 1975). The local loss of material from the surface of the egg is compensated by the extension of the ectodermal moiety towards the lower pole of the egg. Both invagination and epiboly are essentially autonomous processes, as shown by Holtfreter's isolation experiments (1938a, 1943, 1944), but in normal development they are so well coordinated that the egg surface remains smooth during the entire gastrulation process. The invaginating mesoderm accumulates on the dorsal side as the result of dorsal convergence and ventral divergence in the mesodermal mantle (fig. 2.4). These processes go hand in hand with a pronounced cranio-caudal stretching of the dorsal mesoderm. The roof of the archenteron consists chiefly of mesoderm, while its floor consists of yolk-rich endoderm. The archenteron undergoes a process of remodelling in which the two components segregate from each other. The former gives rise to a continuous mesodermal sheet in apposition to the ectoderm, while the latter, after its lateral edges have curved upwards and closed dorsally, forms a tube-like endodermal layer (Vogt, 1929). As a consequence of the invagination process regions of the embryo which were initially far apart come into temporary or permanent apposition, so that new interactions occur, of which the formation of the neural anlage is the most conspicuous one.

We have seen that mesoderm formation is not yet completed by the time the vegetal yolk mass has lost its inductive capacity.

It is known that in the *dorsal* mesendoderm the differentiation tendencies of both the presumptive cranial and caudal archenteron roof change during the process of invagination. When isolated *before* its invagination and enveloped in competent ectoderm, the presumptive cranial archenteron roof forms notochord and somite tissue and induces rhombencephalic and spinal cord structures. The same material, however, forms pharyngeal endoderm and prechordal mesoderm and induces fore-brain structures when isolated directly *after* its invagination (Okada & Takaya, 1942*a, b*; Okada & Hama, 1943, 1944, 1945; Hama, 1949; Hoessels, 1957). The differentiation tendencies of the presumptive caudal archenteron roof also change. While it shows hardly any mesodermal differentiation tendencies at the early gastrula stage, it acquires the capacity to form notochord and somites when it approaches the dorsal blastoporal lip, where it comes into (temporary) contact with the invaginating presumptive cranial archenteron roof. These data suggest that an interaction may occur between the cranial and caudal portions of the presumptive archenteron roof when they are temporarily apposed during invagination. Recombination experiments by Hiradhar (unpublished) suggest that such an interaction does indeed occur.

When the invaginating cranial archenteron roof comes into contact with the overlying ectoderm, prosencephalic neural differentiation tendencies are evoked in the latter. The neural 'activation' process spreads cranially with the extension of the archenteron. Under the influence of the more caudal portions of the archenteron that invaginate subsequently a second inductive action takes place which is superimposed upon the primary activation process. This has been called 'transformation' (Nieuwkoop *et al.*, 1952) and changes the previously evoked prosencephalic differentiation tendencies into those for mesencephalon and rhombencephalon and spinal cord. Thus two successive induction waves, an activating and a transforming one, move in caudo-cranial direction through the presumptive neurectoderm (Nieuwkoop *et al.*, 1952; Eyal-Giladi, 1954).

Recent experiments by Nieuwkoop & Weyer (1978) confirm observations by Yamada (1939), Kato (1957, 1963) and Muchmore (1964) showing that the interaction between archenteron roof and overlying ectoderm is actually a reciprocal process, in which not only is a neural plate induced in the overlying ectoderm but also the notochordal differentiation tendencies in the underlying archenteron roof are markedly enhanced.

Later interactions between the mesodermal mantle and the underlying endoderm lead to the regional differentiation of the gut; these interactions, which were initially studied by Balinsky (1948), have recently been extensively investigated by Albert (1976). The regional differentiation of the epidermis likewise occurs under the influence of the underlying mesoderm.

Summarising, it may be stated that the establishment of the final pattern of differentiation tendencies in the *dorsal* mesoderm requires at least three successive interactions. These are: (*a*) the initial mesoderm-inducing action of the vegetal endodermal yolk mass spreading into the adjacent equatorial region of the animal moiety; (*b*) the subsequent interaction between the 'cranial' and 'caudal' portions of the presumptive archenteron roof during invagination; and (*c*) the reciprocal interaction between the invaginated archenteron roof and the overlying neurectoderm.

Although little is known about the sequence of events in the lateral and ventral marginal zone, at least two successive interactions seem to occur in both regions. After the initial mesoderm-inducing action of the *lateral* regions of the vegetal yolk mass a second inductive action spreads from the median notochordal anlage laterally through the mesodermal mantle. This inductive action evokes (Yamada, 1938, 1940), or at least enhances (Muchmore, 1951), somite and nephrogenic differentiation tendencies and suppresses ventral mesodermal differentiation tendencies in the lateral mesoderm. Although few studies have been made on the induction of the *ventral* mesoderm, the primary mesoderm-inducing action of the ventral region of the vegetal yolk mass (Boterenbrood & Nieuwkoop, 1973) seems to be succeeded by a secondary action of the ventral endoderm which is responsible for blood cell formation (Nieuwkoop & Sutasurya, unpublished).

Anura

The early development of the anuran egg differs from that of the urodele egg in several respects. The egg is usually more strongly pigmented, except in *Rhacophorus*. The anuran egg possesses a mechanism preventing polyspermy, so that fertilised eggs are normally monospermic. As a consequence the role of the spermatozoon in the process of symmetrisation has been better studied in the anuran egg than in the urodele egg. In the former the penetration of the sperm leads to a rapid extrusion of cortical granules into the perivitelline space, a phenomenon which is probably associated with the prevention of polyspermy.

In an extensive study Ancel & Vintemberger (1948) demonstrated that in *Rana temporaria* the so-called 'grey crescent' is formed opposite the place of sperm entry at the junction of the vegetal yolk mass and the animal ooplasm (shown in fig. 2.5 for *Discoglossus*). Although we do not yet know the causal sequence of events that leads from the penetration of the sperm to the appearance of the grey crescent, there are indications that besides contraction phenomena in the cortical layer of the egg, which lead to changes in the

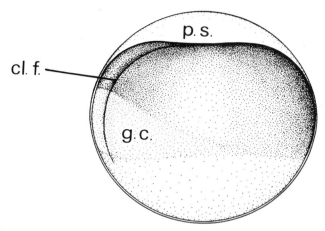

Fig. 2.5. Egg of *Discoglosus pictus* at the beginning of the first cleavage, showing grey crescent (g.c.) bisected by first cleavage furrow (cl.f.). p.s., perivitelline space.

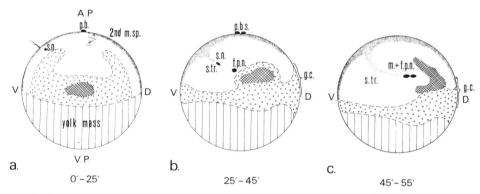

Fig. 2.6. Displacements of internal cytoplasm during the period between fertilisation and grey crescent formation in *Xenopus laevis*. (a) Radially symmetrical configuration, segregation of yolk-poor cytoplasm (hatched) in the region of the egg with medium-sized yolk platelets (stippled); position of first polar body (p.b.), second maturation spindle (2nd m.sp.) and sperm nucleus (s.n.) at 0 to 25 minutes post-fertilisation. (b) Displacements of yolk-poor cytoplasm and cytoplasm with medium-sized yolk platelets towards the dorsal side and the appearance of grey crescent (g.c.); location of female pronucleus (f.p.n.) and tract of sperm nucleus (s.tr.) at 25 to 45 minutes post-fertilisation. p.bs, first and second polar bodies. (c) Final position of yolk-poor cytoplasm and fusion of male and female pronuclei (m.+f.pn.). A.P., animal pole; D., dorsal; V., ventral; V.P., vegetal pole. (After Ubbels, 1978.)

pigment pattern (Hara, unpublished), an internal displacement of cytoplasm towards the dorsal side (Ubbels, 1978) plays a role in the symmetrisation process (fig. 2.6).

In the absence of the directive influence of the sperm – in the case of sperm entrance at the animal pole or of artificial activation – the gravitational rotation of the egg inside its capsules determines the bilateral symmetry. The side along which the vegetal pole of the egg descends becomes the dorsal side. It is also known, however, that in the absence of any directive influence from either the sperm or gravitational rotation, symmetrisation nevertheless occurs. This may be due either to some form of presymmetrisation of the egg during oogenesis or to an intrinsic tendency of the egg to establish bilateral symmetry. As to the latter possibility, perhaps the radially symmetrical configuration of the freshly laid egg represents a labile state of its contractile cortical layer (Nieuwkoop, 1977). Pasteels (1937*a*) claimed that in *Rana esculenta* the eccentric position of the pigment cap with respect to the animal–vegetal axis found in a certain percentage of unfertilised eggs always coincided with the future plane of bilateral symmetry after fertilisation. In these eggs symmetrisation seemed to be independent of the site of sperm entrance.

In the anuran egg the first cleavages occur in approximately the same pattern as in the urodele egg but are less variable, so that after the third cleavage a characteristic 8-cell stage is frequently formed (fig. 2.7). In typical cases the slightly smaller dorsal micromeres are only lightly pigmented and comprise the entire grey crescent region, while the ventral micromeres are strongly pigmented. The dorsal macromeres are completely unpigmented and consist of yolk endoderm only, while the ventral macromeres also include a pigmented region which in our experience forms part of the animal moiety (see p. 11).

During cleavage both the urodele and the anuran embryo are fragmented into a logarithmically increasing number of blastomeres, but in the anuran embryo the cleavage planes are not always perpendicular to the surface. From the sixth to the tenth cleavage some of the cleavage planes are oriented parallel to the surface of the embryo. As a consequence an essentially double-layered blastula is formed with a thin outer layer and an inner layer several cells thick. Gastrulation is preceded by pregastrulation movements similar to those described for the urodele embryo, which lead, among other things, to a fountain-like ascent of individual vegetal blastomeres towards the floor of the blastocoel (cf. the organ anlage map for the 32-cell stage by Takasaki & Yagura, 1975).

The double-layered configuration of the anuran blastula markedly affects the gastrulation process. The outer cell layer forms only the epithelial layer of the epidermis and the ependymal layer of the nervous system, and the endodermal lining of the archenteron. The inner layer gives rise to the

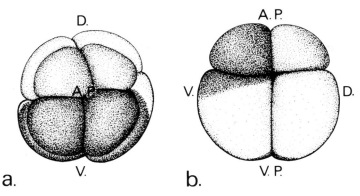

Fig. 2.7. *Xenopus laevis* embryo at the 8-cell stage, with the grey crescent region located in animal dorsal blastomeres and an extension of the animal moiety into the vegetal ventral blastomeres. (a) Animal view, (b) lateral view. A.P., animal pole; D., dorsal; V., ventral; V.P., vegetal pole.

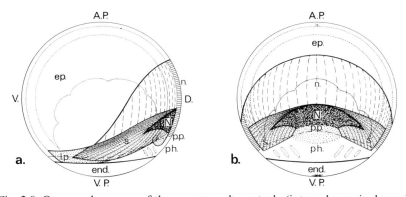

Fig. 2.8. Organ anlage map of the anuran early gastrula (internal marginal zone). (a) Lateral view, (b) dorsal view (slightly tilted). A.P., animal pole; D., dorsal; end., endoderm; ep., epidermis; l.p., lateral plate; n., neural plate; N, notochord; ph., pharynx endoderm; p.p., prechordal plate; s., somites; V., ventral; V.P., vegetal pole. (After Vogt, 1929, as corrected by Pasteels, 1942, and Nieuwkoop & Florschütz, 1950.)

sensorial layer of the epidermis, the bulk of the nervous system, the entire mesoderm and the nutritive yolk mass (see the anuran anlage map, fig. 2.8).

After the initial formation of a dorsal blastoporal groove invagination of the archenteron is temporarily arrested, while the mesoderm continues to roll in around an 'internal blastoporal lip'. Only after a yolk plug has been formed by the ventral completion of the blastoporal groove does a rapid invagination and extension of the purely endodermal archenteron occur, which quickly catches up with the independently invaginating mesoderm (Nieuwkoop &

Florschütz, 1950). These observations have recently been fully confirmed in accurate vital staining experiments by Keller (1975, 1976).

In the anuran embryo neither the mesodermal mantle nor the endodermal archenteron extends up to the animal pole, as they do in the urodele embryo, so that in the former the original animal pole of the egg corresponds to the ventral epidermis of the heart region (Vogt, 1929). This difference between the two groups is also partly due to the fact that the 'internal' marginal zone of the anuran embryo is markedly narrower (45°) than the marginal zone of the urodele embryo (60°) (see also Holtfreter, 1938*b*).

Recombinates of the vegetal yolk mass with the animal hemisphere of *Xenopus* blastulae were analysed qualitatively and quantitatively by Sudarwati & Nieuwkoop (1971). These experiments demonstrated that in the anurans the entire mesoderm is formed exclusively from the internal layer of the animal moiety of the embryo under an inductive influence exerted by the vegetal yolk mass. This inductive action is probably also responsible for the formation of part of the presumptive lining of the archenteron from the external layer of the animal moiety. The results strongly suggest that the process of mesoderm formation in the anuran embryo is nevertheless comparable with that in the urodele embryo.

Coecilia

The Coecilia constitute the third order of the amphibians, which comprises only a small number of species. Their subterranean habitat makes these animals rather inaccessible, so that only a few fragmentary studies of their embryonic development exist.

The egg of *Hypogeophis* forms a blastodisc. Gastrulation starts near the future caudal boundary of the blastodisc with the formation of a dorsal blastoporal lip, which then extends laterally and finally encloses a small endodermal yolk plug (fig. 2.9 a–f). Cranial to the yolk plug an archenteron is formed, which is lined dorsally with chordo-mesoderm and ventrally with endoderm, as in the urodeles. At the anterior end the archenteron seems to communicate with the blastocoelic cavity (fig. 2.9 g, h). The gut is formed by a dorsal fusion of the lateral wings of the endoderm underneath the chordo-mesoderm. A neural plate appears above the archenteron roof around the time of closure of the blastopore. From the periphery of the blastodisc both ectoderm and endoderm gradually extend around the entire yolk mass, forming a yolk sac. During further development the embryonic anlage is lifted up from the yolk sac, to which it ultimately remains connected only by a stalk. After consumption of the yolk the stalk is retracted and the yolk sac incorporated into the rapidly elongating embryo (Brauer, 1897*a*, *b*; Marcus, 1908).

Only this rather fragmentary description of normal development exists.

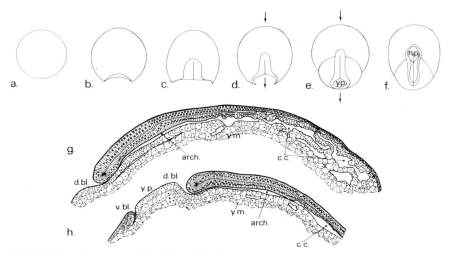

Fig. 2.9. Development of the blastodisc in the embryo of *Hypogeophis alternans* (a–f), and two longitudinal sections (g, h) of stages indicated by arrows in (d) and (e), showing the connection of the archenteron (arch.) with the cleavage cavity (c.c.). d.bl., dorsal blastoporal lip; n., neural plate; v.bl., ventral blastoporal lip; y.m., superficially cleaved yolk mass; y.p., yolk plug. (After Brauer, 1897*b*.)

Brauer's description and illustrations nevertheless suggest that mesoderm formation is of the urodele type.

The fishes

The fishes form a heterogeneous group, of which the extant forms are subdivided into four classes, viz. the very primitive Agnatha, Cyclostomata or lampreys; the Chondrichthyes or cartilaginous fishes; the Osteichthyes, to which a number of ancient forms belong; and the Teleostomi or bony fishes. The Teleostomi comprises the great majority of the present-day fishes and forms a very large and highly specialised group.

On the basis of their embryonic development the fishes may be arranged into two major groups. The first group consists of the mesolecithal Agnatha (the Myxinidae are telolecithal, however) and Osteichthyes, which have an essentially holoblastic cleavage, and the second group the telolecithal Chondrichthyes and Teleostomi, with a meroblastic cleavage.

Agnatha

In *Petromyzon* the processes of cleavage, blastula formation and gastrulation show a pronounced similarity with those of the amphibians (Kupffer, 1890).

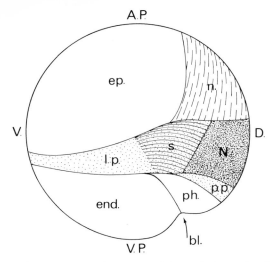

Fig. 2.10. Approximate organ anlage map of the early gastrula of *Petromyzon planeri* (lateral view). A.P., animal pole; bl., blastoporal groove; D., dorsal; end., endoderm; ep., epidermis; l.p., lateral plate; n., neural plate; N, notochord; ph., pharynx endoderm; p.p., prechordal plate; s., somites; V., ventral; V.P., vegetal pole. (After Weissenberg, 1933, as corrected by Daclq, 1935.)

The anlage map made by Weissenberg (1933) and corrected by Dalcq (1935) (fig. 2.10) is essentially homologous with that of the urodeles (cf. fig. 2.3). Early embryogenesis is likewise very similar to that of the amphibians. In the lamprey the yolk mass is finally located in the posterior trunk region, as a result of the early elevation of the head (Kupffer, 1890; Okkelberg, 1921).

Osteichthyes

In *Acipenser*, where fertilisation occurs through the micropyle at the animal pole, symmetrisation becomes externally visible after grey crescent formation and seems to be caused by the gravitational rotation of the egg inside its capsules (fig. 2.11a–c). As in the amphibians the eggs essentially have a holoblastic cleavage, although markedly retarded in the vegetal half (fig. 2.11d). The blastula is essentially single-layered. The process of the invagination of the mesoderm and the formation of the mesodermal mantle is very similar to that in the urodele amphibians (fig. 2.11e–f). In the sturgeon the yolk mass is finally situated in the anterior trunk region, as a result of early and rapid tail bud formation which occurs before the elevation of the head (Ginsburg & Detlaff, 1955; see also Ballard & Needham, 1964, who studied the development of the degenerate form *Polyodon*). Ignatieva (1960, 1962) carried out some experimental studies on the inductive capacity of the

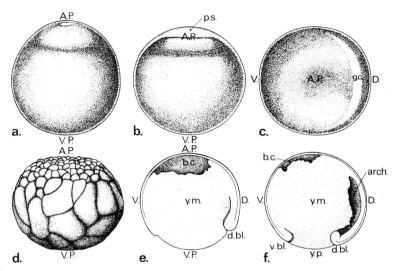

Fig. 2.11. Development of *Acipenser stellatus*. (a) Mature egg in side view. (b) Flattening of animal pole region after activation. (c) Formation of grey crescent (g.c.) in animal view. (d) Early blastula stage. (e) and (f) Median sections of two different stages showing the gastrulation process. A.P., animal pole; arch., archenteron; b.c., blastocoelic cavity; D., dorsal; d.bl., dorsal blastoporal lip; p.s., perivitelline space; V, ventral; v.bl., ventral blastoporal lip; V.P., vegetal pole; y.m., yolk mass; y.p., yolk plug. (After Ginsburg & Dettlaf, 1955.)

chordo-mesoderm in the sturgeon. Neural induction in the sturgeon shows distinct similarity with that in the amphibians, although the differentiation tendencies and inductive power of the dorsal marginal zone are weaker than in the amphibian early gastrula.

The egg of *Lepisosteus*, which is richer in yolk than the sturgeon egg, shows a combination of blastodisc formation and retarded holoblastic cleavage (Balfour & Parker, 1882).

Experimental studies on the origin of the mesoderm have unfortunately been made neither in the lamprey nor in the sturgeon. On the basis of normal development it nevertheless seems likely that in the Osteichthyes mesoderm formation is essentially similar to that in the urodele amphibians.

Chondrichthyes

Among the early workers on the meroblastic Chondrichthyes there was much controversy about the absence or presence of a real gastrulation process. Balfour (1877) and Hoffmann (1896) denied the existence of real invagination in the blastodisc of the shark egg, whereas Ziegler & Ziegler (1892) concluded that the internal layers are formed by invagination. Vandebroek (1936)

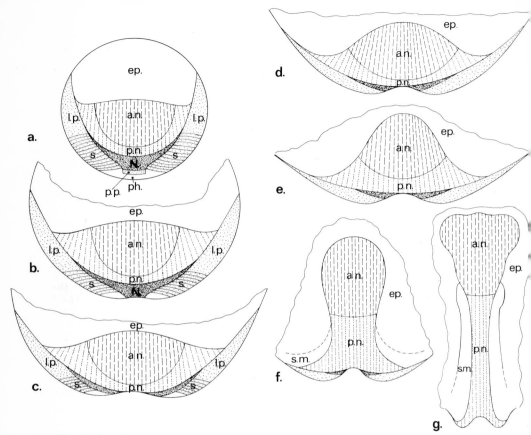

Fig. 2.12. Displacements of organ anlagen during gastrulation and neurulation in *Scyllium canicula*. a.n., anterior neural plate; ep., epidermis; l.p., lateral plate; N., notochord; ph., pharynx endoderm; p.n., posterior neural plate; p.p., prechordal plate; s.m., somitic mesoderm. (After Vandebroek, 1936.)

applied vital staining to the egg of *Scyllium*. He demonstrated that after the initial radial extension of the blastodisc, vitally stained areas actually leave the surface and invaginate around the caudal (= dorsal) edge of the blastodisc. Invagination is, however, restricted to the caudal and lateral edges of the blastodisc. Median stained areas end up in front while more lateral stained areas follow after swinging towards the midline. The invaginated material forms the prechordal plate, notochord and somites. The blastodisc shows strong 'dorsal' convergence movements both in the external layer and in the invaginated material. Simultaneously the embryonic anlage stretches considerably (fig. 2.12). The anlage map of the circular blastodisc shows interesting similarities to that of the amphibian blastula. However, instead of the ring-shaped marginal zone of the amphibians the blastodisc of *Scyllium* seems to have a horseshoe-shaped marginal zone which embraces the more or less circular ecto-neurodermal anlage. The endoderm which at least partially underlies the marginal zone seems to be formed already before gastrulation (Vandebroek, 1936).

During further development the embryo is lifted up from the yolk, with which it is finally connected only by a thin stalk.

No experimental data exist on interactions between the various layers, so that nothing can be said about a possible epigenetic development of the mesoderm.

Teleostomi

The teleost egg shows a rather aberrant development. Shortly before cleavage a so-called 'polar differentiation' of the egg occurs as a result of contraction waves which move over the egg towards the animal pole, lifting up the blastodisc from the yolk mass (e.g. Lewis, 1942; Ballard, 1973*a*) (fig. 2.13a, b). Blastodisc and yolk mass are connected by a thin layer of cytoplasm, the periderm. Symmetrisation of the teleost egg is a slow process which extends over part of the cleavage period. No constant relationship exists between the plane of bilateral symmetry and the first cleavage plane (Oppenheimer, 1936*c*). The causal factors leading to symmetrisation are unknown.

Cleavage occurs only in the blastodisc. This consists of a continuous outer cellular envelope, which is firmly attached to the periderm at its periphery, and a number of internally situated blastomeres (fig. 2.13c, d). During cleavage nuclei enter the periderm, forming a syncytium. This vitelline syncytium has a triple function: (1) in the mobilisation of the yolk resources, (2) as a transitional zone between blastodisc and yolk, and (3) as an adhesive substrate for the epibolic expansion of the germ ring, which develops from the blastodisc (Vakaet, 1950, 1955; Devillers, 1961; Oppenheimer, 1964). The freely moving blastomeres form almost the entire future embryo. The cellular envelope, which gradually extends around the entire yolk mass, contributes neither to the inner germ ring nor to the periderm. It is involved only in the overgrowth of the yolk mass and is at all times separable from the inner blastomeres, and later from the embryo. Although contributing to the outer layer of the future epidermis it is mainly an extra-embryonic structure, which is finally sloughed off during larval development (Bouvet, 1976).

Pasteels (1936) and Devillers (1961) state that in the teleosts, as in the sharks, the presumptive notochordal and somite mesoderm invaginates around a dorsal blastoporal lip. Gamo (1961*a*) speaks of 'poly-invagination'. In a careful study involving vital staining of both the upper and the deeper surface of the blastodisc and the injection of chalk particles between the inner

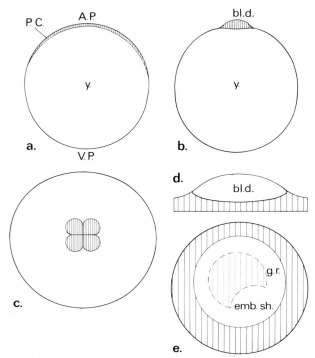

Fig. 2.13. Development of *Salmo gairdneri*. (a) Mature egg with peripheral cytoplasm (p.c.). (b) Formation of blastodisc (bl.d.) on the top of yolk (y). (c) 4-cell stage. (d) Multi-layered blastodisc. (e) Formation of germ ring (g.r.) with embryonic shield (emb.sh.). A.P., animal pole; V.P., vegetal pole. (Drawn from photographs by Ballard, 1973*a*.)

blastomeres, Ballard (1964, 1965, 1966*a*, *b*, *c*, 1968, 1973*a*, *b*, *c*, *d*) and Ballard & Dodes (1968) come to the conclusion that no invagination of superficial material occurs in the teleost embryo. Instead of moving centripetally away from the blastodisc rim, the cells of the deep central part of the blastodisc move outwards towards the rim of the blastodisc and towards the prospective axis. The hypoblast is similarly formed by outward migration of deep central cells. Ballard concludes that there exists a three-dimensional organisation in the blastodisc at the 'pregastrula' stage, in which the presumptive somite mesoderm is situated on top of the presumptive notochord and endoderm. The presumptive neural material overlaps the material of the first 10 pairs of somites, while finally all this is overlain by presumptive epidermis (fig. 2.14). In the teleosts no morphogenetic movements occur which are comparable with blastopore formation and invagination in the amphibians (or with primitive streak formation in the higher vertebrates). The only morphogenetic

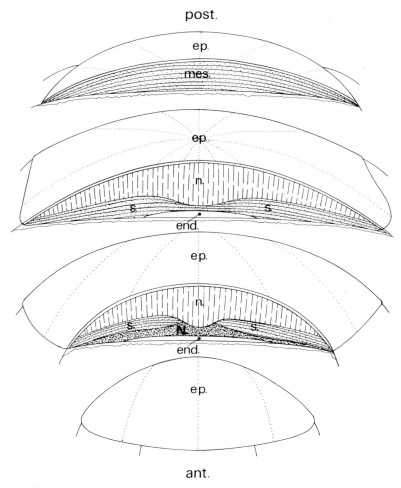

Fig. 2.14. Three-dimensional anlage map of stage 7 blastodisc of *Salmo gairdneri*, divided into four transverse slices. ant., anterior; end., endoderm; ep., epidermis; mes., mesoderm; n., neural anlage; N, notochord; post., posterior; s., somites. (After Ballard, 1973*c*.)

movements are dorsal convergence and stretching movements acting in both the mesoderm and the ecto-neuroderm and directed towards the embryonic axis.

The results of these detailed studies have been laid down in a three-dimensional organ map (Ballard, 1973*c*; fig. 2.15). Such a spatial organisation of the blastodisc at 'pregastrula' stages suggests to us either a spatial segregation of different types of blastomeres during early development, or the

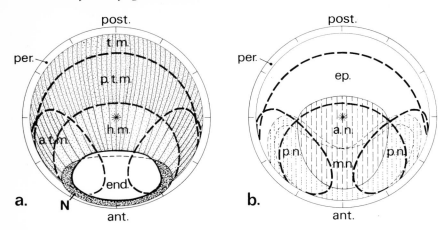

Fig. 2.15. Organ anlage map of stage 7 blastodisc of *Salmo gairdneri*. (a) Seen from below, showing superposition of endoderm (end.), notochord (N, densely stippled) and various regions of mesoderm. a.t.m., anterior trunk mesoderm; h.m., head mesoderm; p.t.m., posterior trunk mesoderm; t.m., tail mesoderm. (b) Seen from above, showing superposition of anterior (a.n.), middle (m.n.) and posterior (p.n.) neural anlagen and underlying mesoderm. *, animal pole; ant., anterior; ep., epidermis; per., periderm; post., posterior. (After Ballard, 1973*c*).

occurrence of inductive interactions between different cell layers of the embryo leading to the formation of the mesoderm and the nervous system, or both.

The teleost embryo is characterised by the presence of a germ ring which is formed by centrifugal migration of inner cells from the outer two-thirds of the blastodisc (underneath the cellular envelope). The embryonic shield is subsequently formed by dorsal convergence in the germ ring as well as by a massive descent of cells from the central third of the blastodisc (cf. fig. 2.13e). Also the hypoblast of the germ ring converges towards the embryonic shield. During the overgrowth of the spherical yolk the germ ring temporarily expands until it occupies an equatorial position, after which it contracts, mainly by dorsal convergence (fig. 2.16). The first sign of embryonic shield formation in the germ ring is a local attachment of cells to the yolk syncytium. These cells probably represent the future prechordal plate cells (see also Knight, 1963).

Although organogenesis clearly proceeds in a cephalo-caudal direction, the caudal part of the embryo being formed much later than the cranial part, it seems incorrect to compare the overgrowth of the yolk by the germ ring with the gastrulation process, and the final closure of the germ ring with the formation of the prospective anus in the other Anamnia. Firstly there is no real invagination process in the teleosts, and secondly the morphogenetic movements leading to embryogenesis have essentially ceased long before the overgrowth of the yolk is completed.

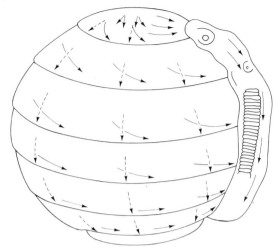

Fig. 2.16. Diagram showing successive steps in epibolic overgrowth of yolk (dashed arrows) and in dorsal convergence in the descending germ ring (solid arrows) of the teleost egg. (After Ballard, 1973c.)

Separation of blastomeres at the 2-cell stage demonstrates that dorso-ventral polarity is localised in the syncytial periderm, in which a crescent-shaped region is visible (Oppenheimer, 1936a; Tung et al., 1945). Even at blastula stages the isolated blastodisc itself has no developmental capacities and only forms a so-called hyperblastula. Only when left *in situ* upon the periderm and the yolk does the blastodisc form an embryo with axial mesodermal structures (Oppenheimer, 1947). Replacement of the ventral half of a blastodisc by a dorsal half at the 'early gastrula' stage leads to double embryo formation. In the reciprocal experiment, in which a ventral half is grafted in place of the dorsal half upon the intact periderm, a normal embryo is formed on the original dorsal side (Luther, 1938). These experiments indicate that in the teleosts dorso-ventral polarity is transferred from the syncytial periderm to the blastodisc. It seems that this transfer takes place during a phase of development which is comparable with gastrulation in the amphibians. The question arises as to whether the syncytial periderm of the teleost embryo could be functionally homologous to the endodermal yolk mass of the amphibian embryo. It is known that the yolk syncytium contributes to the embryonic shield by budding off cells (Devillers, 1961), but it is not known whether these cells contribute to the definitive endoderm of the embryo.

Experiments on later stages show that the developmental capacity for axial structures, which is at first equally distributed over the entire germ ring ('early gastrula' stage), gradually becomes restricted (during 'gastrula' stages) to the embryonic shield which appears in the dorsal portion of the germ ring; the loss of developmental capacity for axial structures starts on the ventral side. The definitive determination of the embryonic shield apparently occurs between a late 'gastrula' stage and an early neurula stage (Luther, 1937). Oppenheimer (1936b) showed that in *Fundulus* transplantation of the dorsal portion of the germ ring of an 'early gastrula' to other embryonic or extra-embryonic regions leads to the formation of an extra axis, so that here the dorsal region of the germ ring already has a certain autonomy. Oppenheimer (1938) concludes that there is a gradation of developmental capacities inside the germ ring due to a directive, inhibitory influence exerted by the dorsal region. It seems more likely, however, that the directive influence is a stimulatory rather than an inhibitory one, since the ventral portion of the germ ring, which is farthest away from the dorsal region, is the first to lose its developmental capacities.

Finally, transplantation of various regions of the embryonic shield led to the conclusion that at the 'mid-gastrula' stage a cranio-caudal axis is gradually established in the embryonic shield, perpendicular to the germ ring (Oppenheimer, 1955, 1959a, b).

Although mesoderm formation has not been explicitly studied in the teleosts, so that we have no proof of an epigenetic development of the mesoderm, the experimental data described above are certainly not inconsistent with, and in fact support, such an hypothesis.

The lower chordates

Cephalochordata

After the penetration of the sperm near the vegetal pole of the egg of *Branchiostoma lanceolatum*, a peripheral layer of cytoplasm accumulates on the lower side of the egg and moves upwards along the 'postero-ventral' side, forming the so-called mesodermal crescent. A second crescent forms on the 'antero-dorsal' side, where the notochord and neural plate anlagen develop. The two crescents are separated on the vegetal side of the egg by the endodermal yolk mass, and on the animal side by an animal cytoplasmic region from which the epidermal structures develop (Conklin, 1932; fig. 2.17a).

According to Conklin the first cleavage plane coincides with the plane of bilateral symmetry, while the second cleavage divides the egg into antero-dorsal and postero-ventral quadrants. The third, equatorial cleavage separates the four animal micromeres from the four vegetal macromeres (fig. 2.18). A cleavage cavity begins to form during the third cleavage and gradually develops into a blastocoelic cavity filled with jelly. From the seventh cleavage onwards (128-cell stage) cleavages are no longer synchronous, the mesodermal cells dividing more rapidly and the endodermal cells more slowly than the other blastomeres. (See also Hatschek, 1881.)

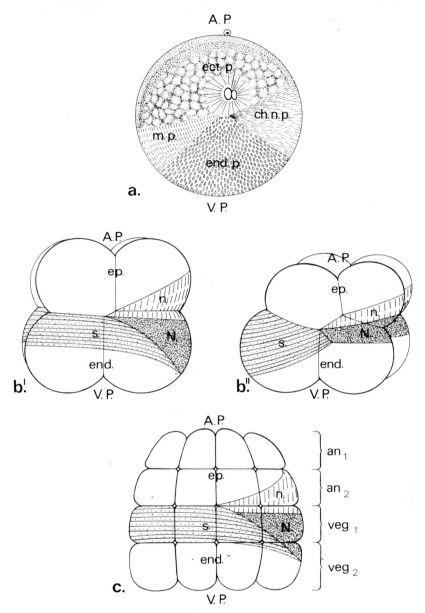

Fig. 2.17. (a) Cytoplasmic segregation in the fertilised, uncleaved egg of *Branchiostoma lanceolatum.* ch.n.p., chordo-neural cytoplasm; ect.p., ectoplasm; end.p., endoplasm; m.p., mesoplasm (after Conklin, 1932). Location of organ anlagen at the 8-cell stage in (b[1]) *Branchiostoma belcheri* after Tung *et al.* (1962*a*), (b[11]) in *B. lanceolatum* after Conklin (1932). (c) Location of organ anlagen at the 32-cell stage in *B. belcheri* after Tung *et al.* (1962*a*). A.P., animal pole; end., endoderm; ep., epidermis; n., neural anlage; N., notochord; s., somites; V.P., vegetal pole; an[1], an[2], veg[1] and veg[2], successive tiers of blastomeres.

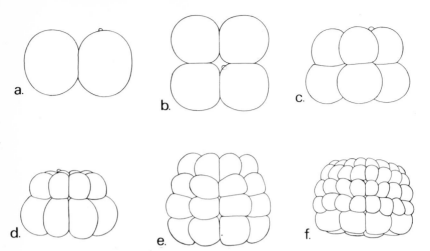

Fig. 2.18. Cleavage stages of *Branchiostoma* up to the 64-cell stage. (After Hatschek, 1881.)

The localisation of the presumptive anlagen was first described by Conklin (1932) for *B. lanceolatum* (fig. 2. 17b[11]) and slightly modified by Tung, Wu & Tung (1962a) for *B. belcheri* (fig. 2.17b[1]). The latter authors also give the localisation of the various anlagen at the 32-cell stage for *B. belcheri* (fig. 2.17c), when the egg consists of four tiers of eight blastomeres, called respectively an_1, an_2, veg_1 and veg_2.

The topography of the presumptive organ anlagen in the *Branchiostoma* embryo shows a striking similarity to that in the ascidian embryo (cf. fig. 2.20) as well as to that in amphibians (cf. figs. 2.3 and 2.8). In all three groups the anlage plan essentially consists of three successive zones: an animal (ecto-neurodermal), and equatorial (mesodermal), and a vegetal (endodermal) region. In the ascidians the neural anlage is predominantly located in the anterior vegetal blastomeres, in *Branchiostoma* in the anterior animal blastomeres. Cytologically the presumptive neural region strongly resembles the presumptive ectodermal one, while the presumptive mesodermal and particularly the presumptive notochordal regions resemble the presumptive endodermal one, although they contain less yolk.

Gastrulation begins after about eight cleavage cycles, when the embryo consists of 256 cells. The blastula flattens dorso-ventrally and becomes pear-shaped as a result of a sinking-in of the mesodermal cells (fig. 2.19a, b). Slightly later the notochordal crescent follows, so that a dorsal blastoporal lip is formed with three transverse inner rows of presumptive notochordal cells and three outer rows of presumptive neural cells. Subsequently cells of the mesodermal crescent become laterally inflected into two lateral horns forming

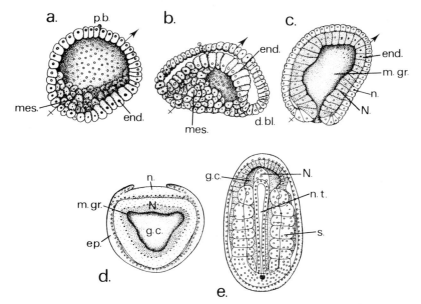

Fig. 2.19. Normal development of *Branchiostoma lanceolatum*. Median sections of (a) early gastrula at 5½ hours of development, (b) middle gastrula at 9 hours, (c) late gastrula at 14 hours, (d) transverse section of neurula at 16 hours, and (e) frontal section of early larva at 18 hours of development. Arrows, future antero-posterior axis; d.bl., dorsal blastoporal lip; end., endoderm; ep., epidermis; g.c., gastrocoel; mes., mesoblast; m.gr., mesodermal groove; n., neural plate; N, notochord; n.t., neural tube; p.b., polar body; s., somites. (After Conklin, 1932.)

the left and right mesodermal pouches, while the crescent's middle region forms the small ventral blastoporal lip (fig. 2. 19b, c). These changes in the mesodermal crescent, which are comparable with the ventral divergence movements of the mesoderm in amphibians, lead to the formation of the mesodermal grooves on either side of the notochordal plate. Subsequently the transversely oriented notochordal and neural anlagen drastically change form by an interdigitation of their cells and become elongated antero-posteriorly. Simultaneously the mesodermal grooves constrict into segmental pouches and separate from the gut, while the ectoderm overgrows the neural anlagen from both sides, closing in from behind (fig. 2.19d, e).

Although the gastrulation process in *Branchiostoma* differs in several respects from that in the amphibians, the same processes of epiboly and invagination, of dorsal convergence and ventral divergence, and of stretching occur in both groups, leading to a more or less identical configuration of the main organ anlagen in the triple-layered embryo.

The first experimental studies were carried out by Conklin in 1933, when he separated various blastomeres. He made the interesting observation that isolated blastomeres of the 2- or 4-cell stage show a cleavage pattern which is identical with that of the whole egg, i.e. the first two cleavages are equal and meridional and the third cleavage unequal and equatorial. Right or left 1/2 or 2/4 blastomeres form normal embryos, but anterior or posterior 2/4 blastomeres only form partial embryos which are complementary in structure. Their differentiation is in accordance with the anlage map. Incomplete separation of blastomeres leads to partial duplication; for instance, into separate anterior regions connected by a single posterior one. Conklin concluded that complete regulation only occurs bilaterally, as a result of redistribution of organ-forming substances already present, and that no regulation occurs in antero-dorsal or postero-ventral blastomeres, which contain different organ-forming substances. Thus, according to Conklin, the *Branchiostoma* egg is essentially a 'mosaic' egg.

Tung, Wu & Tung (1958) repeated the blastomere isolation experiments at the 4-cell stage. In the majority of cases the embryos developed according to the anlage map, as in Conklin's experiments, showing a regulation of symmetrical halves into complete embryos. In a number of cases, however, aberrations were found which clearly showed that the first cleavage plane does not always coincide with the plane of bilateral symmetry and in some cases can even be perpendicular to it.

Isolated animal 'halves' of 8- or 16-cell stages developed into ciliated blastulae, a small proportion of which invaginated, forming a small gut with or without muscle fibres. Isolated vegetal 'halves' gastrulated in the majority of cases and differentiated into notochord, neural tissue, somites and gut, but no ectoderm. The authors concluded that any blastomere containing all five organ-forming regions may regulate into a complete embryo. Although their results are largely in accordance with those of Conklin they nevertheless disagree with this conclusion that the *Branchiostoma* egg is a typical mosaic egg. They found clear indications of interaction between cells, particularly in the formation of the musculature and the nervous system and in the development of the ectodermal tail.

Isolation of individual tiers of blastomeres of the 32-cell stage, gave the following results (cf. fig. 2.17). an_1 formed only atypical epidermis and sometimes tail, an_2 formed epidermis and some mesodermal tissue (in two-thirds of the cases), veg_1 differentiated chiefly into notochord and muscle tissue and occasionally formed some ectodermal or endodermal structures, and veg_2 formed only endodermal cells. These endodermal cells, however, did not differentiate into typical endodermal structures (Tung *et al.*, 1959).

Recombinates of $an_2 + veg_1 + veg_2$, $an_1 + veg_1 + veg_2$, $an_1 + an_2 + veg_2$, and $an_1 + an_2 + veg_1$ formed more or less perfect larvae containing all the essential organ systems. The recombinates $an_1 + veg_2$ and $an_2 + veg_2$ also formed more or less complete larvae. Although each tier of blastomeres has its own developmental capacity, the recombinates show that the *Branchiostoma* egg is endowed with considerable regulative capacity. Complete regulation to half-size embryos can even occur in the $an_1 + veg_2$ recombinates. From this Tung *et al.* (1960*a*) concluded that veg_2 is capable of giving rise to notochord and somites, and that some stimulus is exerted by the an_1 blastomeres towards notochord and somite differentiation. The notochord in its turn exerts an influence upon the ectoderm leading to neural differentiation.

Translocation of the animal blastomeres through $90°$ or $180°$ with respect to the vegetal blastomeres can lead to normal larvae. In such embryos the nervous system is not derived from the presumptive neural material, which forms epidermis, but from the presumptive lateral or ventral ectoderm. After rotation through $180°$ the tail is formed from the anterior blastomeres and not from the posterior ones, while the direction of the ciliary beat is also reversed. This shows that the antero-posterior polarity is determined by the vegetal blastomeres (Tung *et al.*, 1960*b*). Von Ubisch (1963), however, who still accepted the mosaic nature of the *Branchiostoma* egg, ascribed the outcome of Tung's translocation experiments to the slightly varying composition of the individual blastomeres.

At the 32- or 64-cell stage grafts of ectodermal cells of an_1 or an_2 into veg_2 led to an endodermisation of the graft, while grafts of endodermal cells of veg_2 into an_1 invaginated and usually formed only endodermal structures (Tung *et al.*, 1961). Occasionally the invaginated endodermal cells were accompanied by notochord, neural tube and two rows of somites (formation of a secondary axis system). In some cases the endodermal graft was no longer identifiable and seemed to be converted into epidermis. The authors concluded that the ectoderm has great regulative capacity and is readily converted into endodermal cells. The developmental capacity of the endoderm is usually restricted to its prospective significance, and seems to be broader only in exceptional cases. We feel, however, that the last conclusion is not very firmly based.

Implantation of a dorsal blastoporal lip (presumptive notochord) into the blastocoelic cavity of a blastula or early gastrula led to the induction of an extra neural tube in the overlying ectoderm (Tung *et al.*, 1962*b*). No neural induction occurred after implantation of somitic mesoderm or endoderm. Implantation of presumptive notochord in addition led to the formation of two rows of somites. The authors concluded that the presumptive notochord possesses the power of neural induction, and further that the endoderm shows a much higher degree of developmental independence than the chordomesoderm, while the ectodermal tissue is the least independent in its development.

The regulative capacity of the endoderm was investigated by Y. F. Y. Tung *et al.* (1965) by removing a progressively larger mass of vegetal cells from blastula or early gastrula. Removal of one-fifth of the embryo led to normal (30%) or slightly deficient (70%) endodermal structures. After removal of a quarter to one-third of the embryo the endodermal structures were severely deficient (67%) or completely absent (33% of the cases). They concluded that at these stages the endoderm still has a great capacity for internal regulation.

From recombinates of varying numbers (from one to eight) of animal and vegetal blastomeres of the 8- or 16-cell stage Tung *et al.* (1965) concluded that a 1/8 vegetal blastomere, when combined with four to eight animal blastomeres, is still capable of organising an harmonious embryo, provided the vegetal blastomere is an anterior one that includes presumptive noto-chordal material. Perfect regulation does not occur, however, when the total mass is too small or when there is too little presumptive chordo-mesodermal and endodermal material.

These results, which were briefly reviewed by Reverberi (1971*b*), show a striking similarity with those of the recombination experiments carried out by Nieuwkoop and co-workers on amphibian material (cf. review by Nieuw-koop, 1973). The two groups seem to have in common the view that the mesoderm is formed epigenetically through an interaction of vegetal and animal material. Although the Chinese authors assume that notochord and somites are formed from the vegetal blastomeres under an inductive influence from the animal material, no actual proof exists for this. Their results can be satisfactorily explained by assuming that in *Branchiostoma*, as in the amphibians, the mesodermal structures are induced in the totipotent animal material by the vegetal yolk endoderm, the original boundary between the two moieties at the 8- or 16-cell stage being situated below the level of the third cleavage plane, i.e. in the vegetal blastomeres. Viewed in this light all the results in *Branchiostoma* fall perfectly in line with those from the isolation and recombination experiments carried out on the amphibian embryo and with the earlier studies on mesodermal differentiation (e.g. Yamada, 1938, 1940) and neural induction (see Saxén & Toivonen, 1962). On this premise the following general statement can be formulated. In the lower chordates the dorso-ventral polarity resides in the vegetal yolk mass of the blastula, whence it is transferred to the totipotent animal material through the regional induction of the mesoderm, which in its turn is responsible for the regional induction of the nervous system and the regional differentiation of the ectodermal and endodermal layers.

According to Tung *et al.*, in the *Branchiostoma* blastula the vegetal endoderm has 'a much greater degree of developmental independence' than the animal ectoderm, or in other words, is more strongly determined. This is in full accordance with the situation in the amphibian blastula, which consists of a totipotent animal 'ectodermal' cap and an already firmly determined vegetal endodermal yolk mass. The striking similarity between the two groups argues more strongly for the interpretation given above than for that of the Chinese authors. It must however be realised that alongside these two alternative interpretations a third possibility exists, which cannot yet be completely excluded, i.e. that the mesoderm is formed from both the ectodermal and endodermal components by way of the kind of mutual interaction implicit in the double gradient hypothesis proposed by Runnström (1967) for sea urchin development. It therefore seems highly desirable that the crucial recombination experiments as carried out by Nieuwkoop & Ubbels (1972) on the urodele and by Sudarwati & Nieuwkoop (1971) on the anuran blastula should also be carried out on the blastula of *Branchiostoma* to elucidate the real origin of the mesoderm in the Cephalochordata. This may be achieved either by using two species with different species-specific cellular characteristics or by radio-active marking of the animal or vegetal component.

Tunicata

The subphylum Tunicata is subdivided into three classes: the solitary and usually sedentary Ascidia, the pelagic, solitary or colony-forming Thaliacea, and the neotenic Larvacea. Here we will briefly discuss only the development of the ascidians, which has been reasonably well studied in comparison with that of the other groups (see Reverberi, 1971a; Berrill, 1975).

The unfertilised ascidian egg shows no visible architecture and apparently as yet lacks any differential distribution of subcellular components, since normal mini-larvae can develop from meridional, equatorial and oblique halves of unfertilised eggs after subsequent fertilisation (Ortolani, 1958; Reverberi, 1961; Reverberi & Ortolani, 1962). This conclusion is corroborated by experiments in which one-third of the egg cytoplasm was removed from the unfertilised egg without affecting normal development. Likewise, the fusion of two naked unfertilised eggs can lead to the formation of a single normal giant larva (Farinella-Ferruzza & Reverberi, 1969).

After fertilisation, which usually takes place at the vegetal pole, ooplasmic segregation occurs. In *Cynthia* a 'yellow crescent' is formed which runs half-way round the egg in the posterior equatorial region and represents the presumptive muscle tissue, while a 'grey crescent' representing the chordo-plasm and neuroplasm is formed on the opposite side. The two crescents are separated by a vegetal greyish yolky cytoplasm and an animal clear cytoplasm (Conklin, 1905b) (fig. 2.20a). The yellow cytoplasm contains numerous mitochondria and some pigment granules.

The first cleavage is meridional and follows the plane of bilateral symmetry. The second cleavage is also meridional and divides the egg into two posterior blastomeres containing the entire yellow crescent, and two anterior blasto-

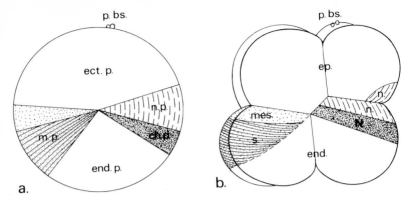

Fig. 2.20. (a) Cytoplasmic segregation in the fertilised, uncleaved egg of *Cynthia*. ch.p., chordoplasm; ect.p., ectoplasm; end.p., endoplasm; m.p., mesoplasm; n.p., neuroplasm; p.bs, polar bodies. (Redrawn from Conklin, 1905*b*.) (b) Location of organ anlagen at the 8-cell stage of an Ascidian. end., endoderm; ep., epidermis; mes., mesenchyme; n., neural plate; N., notochord; s., somites. (After Ortolani, 1954.)

meres containing the grey crescent. The third cleavage is equatorial and divides the egg into four animal and four vegetal cells. At the 8-cell stage the anterior animal blastomeres contain part of the anlagen of the nervous system and the epidermis. The posterior animal blastomeres contain the rest of the epidermal anlage. The anterior vegetal blastomeres contain the rest of the anlage of the nervous system, the entire notochordal anlage, and part of the anlage of the endoderm. The posterior vegetal blastomeres contain the entire anlage of the mesoderm and the remainder of the anlage of the endoderm (fig. 2.20b). Isolated blastomeres of the 2- or 4-cell stage as a rule differentiate according to their prospective significance, although regulation of symmetrical halves into normal mini-larvae occurs (Conklin, 1905*a*; Vandebroek, 1936, see Dalcq, 1938; Ortolani, 1954; Reverberi, 1971*a*).

Farinella-Ferruzza & Reverberi (1969) claim that a normal giant larva may result from fusion of two 2-cell stages, provided the morphogenetic movements of the two embryos are congruous. Reverberi & Gorgone (1962) even obtained a normal larva after fusion of two 8-cell stages by pressing homologous regions together.

Isolated blastomeres of the 8-cell stage developed according to their prospective significance, except that neural differentiation only occurred in the combination of anterior animal with anterior vegetal blastomeres. Replacement of anterior animal by posterior animal blastomeres at 8- to 64-cell stages led to the absence of neural differentiation (Farinella-Ferruzza, 1959; Ortolani, 1959). Ortolani believes that the evocation of the neural anlage occurs at about the 128-cell stage, when notochordal cells and anterior endodermal cells have invaginated and have come into contact with the ectoderm. Von Ubisch (1951, 1952, 1963), however, claims that normal brain and palp formation also occur after destruction of either the endoderm, the mesoderm or the notochord. He infers autonomous development of the neural structures as well as the palps. According to Reverberi (1971*a*), von Ubisch failed to remove all the notochordal cells. Nevertheless Reverberi states that in the ascidians the process of brain formation is more evocative (permissive) than truly inductive (instructive). The presumptive neural material, already being strongly conditioned for neural differentiation, requires only a weak inductive stimulus for its ultimate differentiation, while the inductive stimulus is apparently too weak to evoke neural differentiation in the less competent posterior ectoderm. Reverberi claims that both the notochord and the anterior endoderm play a role in this evocation process (cf. also Rose, 1939; Vandebroek, 1938, 1961).

Gastrulation occurs as early as between the sixth and the seventh cleavage, that is, between the 64- and 128-cell stage. During the invagination process the notochordal crescent converges, and the cells interdigitate and form a single elongated cell mass, which comes to lie between the two diverging arms of the posterior mesoderm. A similar convergence of the neural crescent leads to the formation of the neural plate with its anterior sensory vesicle. The trunk and tail grow out after further interdigitation of the notochordal cells and segmentation of the mesoderm into two rows of somites (Conklin, 1905*a*; Ortolani, 1955, 1971). The gastrulation and neurulation processes in ascidians and cephalochordates show great similarity; in both groups previously separated areas are brought into juxtaposition.

It is very unfortunate that nothing is known about a possible epigenetic development of the mesoderm in the Ascidia. The fact that in the ascidians gastrulation occurs early, between the sixth and seventh cleavage, as against between the eighth and tenth cleavage in *Branchiostoma* and between the fifteenth and eighteenth cleavage in amphibians, suggests that in the ascidians mesoderm formation may take place at an earlier stage of development than in *Branchiostoma*, so that it would practically elude experimental analysis. On the basis of our present knowledge it seems evident that the ascidian egg develops according to a rather rigid pattern from the 8-cell stage onwards, leaving little room for epigenesis during later development.

The Amniota

Among the Amniota most experimental data about mesoderm formation have been obtained from the analysis of early development in the Aves, which will therefore be discussed first. We will then discuss the early development of the Mammalia and conclude with the development of the Reptilia. Since the Reptilia show features which are transitional between the lower and higher vertebrates it might seem more logical to treat the reptiles first, but unfortunately too little is known about them.

Aves

The avian egg is fertilised in the oviduct. During its passage through the oviduct and the uterus the egg becomes surrounded by layers of albumen, the shell membrane, and finally the calcareous shell. The egg starts development immediately upon fertilisation, and during its 20–24 hours' stay in the uterus reaches a single-layered blastoderm stage comparable with an advanced blastula stage in the Anamnia. Kochav & Eyal-Giladi (1971) have shown that the symmetrisation of the hen's egg occurs during the second half of uterine life, more precisely between the fourteenth and sixteenth hour. During passage through the uterus the egg slowly turns on its longitudinal axis (Clavert, 1961). Gravity tends to keep the yolk in a position with the blastoderm uppermost. This tendency is slightly counteracted by the surrounding albumen and by the chalazae by which the yolk is attached to the shell membrane and to the air chamber. The resultant of these opposing forces keeps the blastoderm more or less constantly in an oblique position. In this position gravity determines its symmetrisation. This was demonstrated by Kochav & Eyal-Giladi (1971) by hanging the uterine egg by one of its chalazae in a beaker filled with saline: the embryonic axis always developed with the head down. After the sixteenth hour in the uterus the axis can no longer be altered. Since the blunt end of the egg is usually in front as the egg passes through the uterus, the orientation of the blastoderm in general follows von Baer's rule.

Eyal-Giladi & Kochav (1976) describe the intra-uterine development of the hen's egg as occurring in three successive developmental phases: cleavage, formation of the area pellucida, and formation of the primary hypoblast. During the cleavage period the diameter of the blastoderm at first decreases and its thickness increases. The formation of the area pellucida is the result of a massive loss of yolk-laden cells from the lower surface of the blastoderm, a process extending from back to front. The *primary* hypoblast is subsequently formed by a similarly directed splitting-off of cells from the area pellucida, the descending cells coalescing into a continuous cell layer. Wolk & Eyal-Giladi (1977) demonstrated that in contrast to the epiblast the primary hypoblast acquires specific antigenic properties directly after its segregation from the single-layered blastoderm. This segregation process is erroneously called 'poly-invagination' or 'diffuse invagination' (cf. Pasteels, 1937); no real invagination occurs, however. The segregation of the two layers is preceded by an unequal distribution of cellular components in the original single layer; this may be considered as homologous with the polar differentiation of the anamnian egg into two different moieties, an animal and a vegetal one. Whereas this process occurs during oogenesis in the amphibians and shortly before cleavage in the teleosts, in birds it takes place much later in development. Meanwhile it is not clear how far earlier observations made by Fraser (1954)

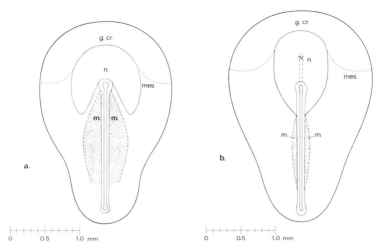

Fig. 2.21. Location of presumptive mesoderm (m., stippled) and neural plate (n.) in chick blastoderm, (a) at definitive primitive streak stage (stage 4, Hamburger & Hamilton, 1951) and (b) at head-process stage (stage 5, H. & H.). g.cr., germinal crescent; mes., anterior border of invaginated mesoderm; N., notochord. (Redrawn from Spratt, 1952.)

and Lutz (1955), suggesting some form of cellular ingression along the periphery of the blastoderm, are correct.

Vakaet (1970) describes the formation of a *secondary* hypoblast from the thickened posterior edge of the blastoderm, the so-called Koller's sickle. The secondary hypoblast pushes the primary one anteriorly. Subsequently a primitive streak is formed, extending forward from the posterior edge of the blastoderm. When this has grown to full length a median groove appears in it, which ends anteriorly in the primitive pit of Henson's node. As soon as groove and pit appear, cells from the upper epiblast layer move towards the midline, turn inwards into the groove and pit region, and then spread laterally between epiblast and hypoblast, forming the mesodermal germ layer. Carbon-marking experiments by Spratt (1946, 1952, 1955, 1957*a, b*) have elucidated the morphogenetic movements in the epiblast layer and the localisation of the presumptive mesodermal and ecto-neurodermal anlagen in the epiblast at the definitive primitive streak (stage 4, Hamburger & Hamilton, 1951) and head process stages (stage 5, H. & H.) (fig. 2.21). Spratt also described the laying down of the notochord and somite anlagen during the regression of the streak and the origin of the extra-embryonic mesoderm from its more posterior section. Spratt & Haas (1960*a, b*, 1965) extended the carbon-marking experiments to the hypoblast by turning the blastoderm upside down and marking its lower surface. They observed a fountain-like outward streaming of cells from a centre near the posterior border of the area pellucida at the

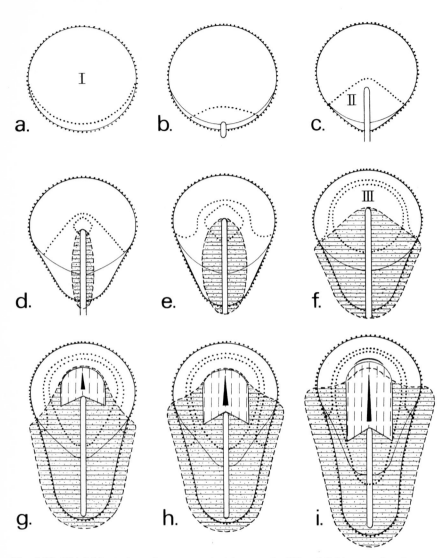

Fig. 2.22. Chick blastoderm from pre-primitive streak till head-fold stage: development of primary (I), secondary (II) and tertiary (III) hypoblast (stippled lines); formation of mesoblast (broken lines and horizontally hatched and stippled) and notochord (black); development of neural anlage in epiblast layer (solid lines and vertically hatched). (Redrawn after Vakaet, 1970.)

pre-streak stage, as well as rapid lateral and anterior movements at the definitive primitive streak stage. Recent cinematographic and radio-active marking experiments by Modak (1966), Rosenquist (1966, 1970, 1971), Vakaet (1970) and Nicolet (1970, 1971) have demonstrated that the first material which invaginates just posterior to Hensen's node is presumptive endoderm which becomes incorporated into the hypoblast layer. This *tertiary* or *definitive* hypoblast gives rise to the entire embryonic endoderm, so that both the primary and the secondary hypoblast ultimately form extra-embryonic endoderm only (fig. 2.22). The primary hypoblast seems to end up in the extra-embryonic region in front of the embryonic anlage, in the so-called 'germinal crescent'. These conclusions were confirmed experimentally by Fontaine & Le Douarin (1977) in quail-chick chimaeras made of epiblasts and hypoblasts of pre-streak stages.

Experiments by Waddington (1932) in which the hypoblast was turned through 90° or 180° with respect to the epiblast gave the first indication that streak formation might be induced by the underlying hypoblast and that the polarity of the embryo might be transferred from the hypoblast to the epiblast. Spratt & Haas (1961) were able to show that the epiblast is not polarised and that its later polarisation depends upon the underlying hypoblast. In contrast, Gallera & Castro-Correia (1964) came to the conclusion that in the unincubated blastoderm the epiblast is already polarised. The recombination experiments on which this conclusion is based seem inadequate, however, since the hypoblast may be re-formed after its removal. The capacity of the hypoblast to induce primitive streak was confirmed by Gallera & Nicolet (1969).

Eyal-Giladi & Spratt (1965) showed that the unincubated blastoderm of 'winter' eggs is almost equipotential, so that even central fragments can form an embryo. Unincubated 'summer' eggs, or 'winter' eggs incubated for $4\frac{1}{2}$ hours lose this embryo-forming capacity in the central portion of the blastoderm; it is now restricted to the marginal zone and gradually becomes localised in its posterior region only. Eyal-Giladi (1969) found that longitudinal folding of the blastoderm had no effect on the orientation of the embryonic axis but that transverse folding strongly influenced the location and orientation of the embryo-forming centre. Removal of the hypoblast from primitive streak blastoderms reduces their developmental capacity to that of unincubated blastoderms, demonstrating the strong directive influence of the hypoblast (Eyal-Giladi, 1970a). Eyal-Giladi & Wolk (1970) concluded from trans-filter induction studies that two successive inductive actions are exerted by the primary (and secondary) hypoblast: induction of a primitive streak and induction of prosencephalic neural structures. We feel, however, that these successive inductive effects may primarily depend upon changes in the competence of the epiblast, which may at first be predominantly mesodermal and later predominantly neural, similar to the situation in amphibian gastrula ectoderm as studied by Leikola (1963).

The inducing capacity of the primary and secondary hypoblast has apparently disappeared by the primitive streak stage (Gallera & Nicolet, 1969; Eyal-Giladi, 1970b) and the definitive embryonic hypoblast induces only during its invagination. Vakaet (1964, 1965) showed that the anterior portion of the young primitive streak, when implanted into the periphery of the area pellucida, can induce brain, while the middle portion induces only a new primitive streak. A graft of Hensen's node from a medium primitive streak stage usually forms embryonic endoderm and induces either brain or a new primitive streak; when it forms axial mesoderm it induces only neural structures.

It may be concluded that in birds all the mesodermal structures as well as the tertiary hypoblast are induced in the totipotent epiblast by the underlying primary and secondary hypoblast. The symmetrisation of the blastoderm apparently first becomes manifest in the hypoblast and is transferred to the epiblast through the regional induction of the mesoderm and endoderm. This situation is analogous to that in amphibians, except that in amphibians endoderm induction is restricted to the pharyngeal endoderm, whereas in birds the entire embryonic endoderm is formed epigenetically during primitive streak formation.

Mammalia

Mammalian development will be discussed only briefly, since experimental data are still rather scanty due to the only very recent development of in-vitro techniques that allow experimental analysis of early developmental stages.

Before discussing the Placentalia or eutherian mammals, a few words must be said about the primitive Monotremata and the Marsupialia, which show a rather aberrant early development.

The Monotremata have an intra-uterine development of about 2 weeks, in which the embryo grows from an initial diameter of about 4 mm to a size of 12–15 mm. After oviposition the egg is incubated for about 2 weeks. The egg cleaves meroblastically. Around the blastoderm a syncytial germ ring is formed. The blastoderm, which originally is seven to eight cell layers thick, thins out rapidly into a single layer while spreading over the yolk. The latter becomes completely surrounded even before a primitive streak is formed. In the single-layered blastoderm a segregation process occurs through which an outer epiblast and an inner hypoblast layer are formed, after which the primitive streak is formed (Flynn & Hill, 1947).

The egg of the Marsupialia is much smaller, about 0.25 mm, but larger than the egg cell of the placental mammals. The nucleus is eccentric and half of the egg is occupied by a large yolk vacuole. Around the egg cell a layer of albumen and a shell membrane are formed. The egg divides into two equal blastomeres while the yolk vacuole is extruded. At the 16-cell stage the blastomeres form two tiers of cells which are morphologically distinct, the so-called animal and vegetal cells. During further cleavage the cells spread along the inner surface of the shell membrane and finally form a single-layered blastocyst (inside which the yolk vacuole dissolves) which consists of two distinct moieties. The smaller, animal cells – the so-called 'formative' cells – give rise to the blastoderm while the vegetal cells – the 'non-formative' cells – form the extra-embryonic blastocyst cells, which are probably homologous with the trophoblast cells of the eutherian mammals. Inside the blastoderm, which forms part of the blastocyst wall, a segregation process occurs through which an outer ectodermal (epiblast) layer and an inner endodermal (hypoblast) layer are formed. While a primitive streak appears in the epiblast the hypoblast spreads over the entire inner surface of the blastocyst (Hill, 1910).

According to Vanneman (1917), in the polyembryonic *Armadillo* the 'formative' cells round up, withdraw from the trophoblast and become surrounded by endoderm. Cavitation occurs in the embryonic mass and two primary embryonic anlagen (called primary embryonic buds) are formed on opposite sides. Subsequently each primary bud gives rise to two secondary diverticula, so that finally four primitive streaks develop.

The small egg of the Placentalia, measuring only 60 to 180 μm, contains only very little yolk. It divides into an exponentially increasing number of blastomeres, which are initially fully equipotential. Usually between the 8- and the 16-cell stage the periphery of the cell mass can no longer accommodate all the blastomeres, so that some cells find themselves in the interior. The topographical distinction between peripheral and internal cells most probably constitutes the first cause of differential development. The outer cells become flattened and polygonal in shape and start secreting fluid into the interior. They subsequently differentiate into trophectoderm or trophoblast cells, whereas the cells of the inner cell mass remain totipotent. Trophoblast cells can no longer take part in embryo formation, whereas inner cells are probably still capable of contributing to the outer layer. In the mouse the determination of inner cell mass versus trophoblast occurs between days $2\frac{1}{2}$ and $3\frac{1}{2}$. The embryo arises exclusively from the inner cell mass (Gardner, 1975; Gardner & Papaioannou, 1975; Daniel, 1976; Gardner & Rossant, 1976).

The formation of the trophoblast cavity leads to a local separation of the trophoblast layer (mural trophectoderm, fig. 2.23a–c) from the inner cell mass. Where the inner cell mass is still attached to the outer layer the trophoblast does not develop into ordinary trophoblast cells (polar trophectoderm, fig. 2.23a–c). In the mouse the determination of mural versus polar trophoblast occurs between days $3\frac{1}{2}$ and $4\frac{1}{2}$. From the mural trophectoderm primary, polyploid giant cells develop. Giant cell formation starts at the abembryonic

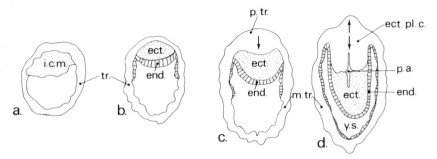

Fig. 2.23. Diagram of development of mammalian (mouse) blastocyst. (a) Segregation of trophoblast (tr.) and inner cell mass (i.c.m.). (b) Segretation of endoderm (end.) and totipotent ectoderm (ect.). (c) Outgrowth of endoderm along inner surface of trophoblast and development of mural (m.tr.) and polar (p.tr.) trophoblast. (d) Formation of yolk sac (y.s.), ecto-placental cone (ect.pl.c.) and proamnion (p.a.). (Redrawn from Gardner & Papaioannou, 1975.)

pole of the trophoblast, suggesting an inhibiting influence emanating from the inner cell mass and spreading with decrement through the trophoblast.

In contrast to the mural trophoblast cells, the polar trophoblast cells at first remain diploid and give rise to a conical cell mass, the 'ecto-placental cone', which in the mouse forms the attachment site of the embryo to the maternal tissues. Attachment is only possible after the zona pellucida has been shed. The mural trophoblast can evoke a decidual response, the inner cell mass cannot (Gardner, 1972). At the periphery of the ecto-placental cone secondary giant cells are later formed. The process spreads laterally until finally the entire conceptus is surrounded.

The next step in development occurs in the inner cell mass itself, where an endodermal cell layer segregates along the free inner surface, leading to a double-layered embryonic anlage (fig. 2.23b). The endoderm subsequently spreads along the inner surface of the trophoblast; as soon as the entire mural trophoblast is lined by 'distal' endoderm it is called yolk sac (fig. 2.23c, d). As in the primitive mammals, the segregation of the endoderm shows distinct similarities with the formation of the primary hypoblast in early avian development. It divides the embryonic anlage into two distinctly different components: an 'ectodermal' mass which is probably still totipotent, and an already determined endodermal layer. This conclusion is supported by recent experiments of Levak-Švajger & Švajger (1970) and Skreb, Švajger & Levak-Švajger (1976), who studied the developmental potentialities of the isolated germ layers at successive stages of development. The epiblast isolated from the embryonic shield of pre-streak and primitive streak stages, when transplanted under the kidney capsule of syngeneic adult rats, gives rise to teratomas containing derivatives of all three germ layers, whereas the hypoblast forms only endodermal derivatives.

In the cone-shaped ectodermal mass a cavity appears (the proamnion), which develops into the amniotic cavity surrounding the future embryo (fig. 2.23d). As a consequence an embryonic shield is formed, which separates the amniotic cavity from the yolk sac. Subsequently a primitive streak develops in the embryonic shield, through which embryonic and extra-embryonic mesoderm invaginates. In the extra-embryonic mesoderm the exocoel or extra-embryonic coelom is formed, which extends all around the amnion, the yolk sac, and the allantois which develops in the caudal portion of the embryonic anlage. The ultimate differentiation of the mesoderm seems to lead to inductive interactions with the underlying endoderm and the overlying ectoderm.

From the primitive streak stage onwards mammalian embryogenesis shows very striking similarities with that of the avian embryo. The mammalian embryo, being very poor in nutritive resources, depends at a very early stage on a food supply from the maternal organism. Blood vessels grow out along the allantoic stalk and ecto-placental cone and, together with the chorion and the allantois, form the chorioallantoic placenta, where exchange with the maternal vascular system is definitively established.

The isolation experiments of Grobstein (1952) and Skreb *et al.* (1976) demonstrate further that the developmental potentialities of the epiblast become restricted at later stages of development. Whereas the epiblast of the primitive streak stage forms teratomas containing derivatives of all three germ layers, the epiblast of the head-fold stage can only form ecto-neurodermal and mesodermal derivatives. This suggests that, as in the avian embryo, the epiblast of the primitive streak embryo may still contain presumptive endodermal as well as mesodermal and ectodermal material, while by the head-fold stage the endodermal cells have left the epiblast. Unfortunately, this is all we know at present about mesoderm formation in the mammalian embryo. It seems plausible, however, that, just as in the birds, primitive streak formation represents mesoderm and endoderm formation induced in the totipotent epiblast by the underlying endoderm (primary hypoblast).

Reptilia

The present-day reptiles are taxonomically subdivided into three orders. According to their development they may be grouped into two different categories, viz. the Chelonia or turtles on the one hand and the Squamata or lizards, skinks and snakes on the other. The Crocodilia, which are certainly more closely related to the lizards than to the turtles, may constitute a third category. Since nothing is known about the early development of the crocodiles we will restrict ourselves to the first two categories.

All reptilian eggs are very rich in yolk and show meroblastic cleavage. A blastoderm is formed on top of the yolk, with a connecting yolk syncytium in between, a situation which on the one hand resembles that in the

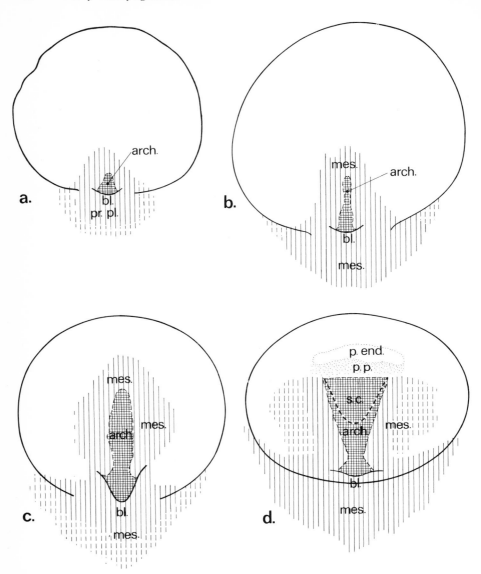

Fig. 2.24. Successive stages (a–d) of gastrulation in the reptile, *Clemmys*; blastopore (bl.) and archenteron (arch.) formation, and invagination of mesoderm (mes.). p.end., prechordal endoderm; pr.pl., primitive plate; p.p., mesodermal prechordal plate; s.c., communication between archenteron and subgerminal cavity. (Redrawn from Pasteels, 1957*a*.)

meroblastic elasmobranchs and teleosts, but on the other hand shows common features with early avian development. In the reptiles, as in the teleosts, cells bud off from the yolk syncytium and contribute to the blastoderm.

Subsequently the blastoderm, which is initially several cell layers thick, begins to expand. Its central portion thins out into a single cell layer – the so-called area pellucida, below which a subgerminal cavity is formed – while the periphery of the disc thickens, particularly on its future posterior side, forming the so-called blastoporal or primitive plate (fig. 2.24).

Although the initial development of the two groups of reptiles is rather similar, their further development seems to differ in various respects, particularly as regards the formation of the primary hypoblast. When studying early chelonian development Pasteels (1937*b*) concluded that the hypoblast is formed by invagination at the postero-lateral margin of the embryonic anlage, which is at variance with the conclusion of Peter (1938), who claims that in the reptiles the primary hypoblast is formed by delamination. Pasteels (1957*b*) makes a sharp distinction between the early development of the chelonians and that of the lacertilians. Although he maintains that in the chelonians both the embryonic and the extra-embryonic hypoblast are formed by cell immigration from the posterior region of the blastoderm, he agrees that in the lacertilians the primary hypoblast is formed *in situ* by delamination. In both groups the primary hypoblast is formed before the blastopore appears. According to Hubert (1976) the primary hypoblast is well developed in *Lacerta*, less well developed in *Vipera*, and extremely reduced in the chelonians. In the chelonians the definitive hypoblast develops during gastrulation.

In the chelonians a sickle-shaped, posteriorly convex blastopore is formed which later changes into a transverse slit and finally becomes concave. In the lacertilians a transverse blastoporal groove is formed, which subsequently becomes concave, later changes into a horseshoe-shaped blastopore, and finally becomes circular. This 'yolk-plug-like' protrusion is not homologous with the amphibian endodermal yolk plug, however, but represents a thickened portion of the ventral blastoporal lip.

Gastrulation proceeds in a similar way in both groups. The prechordal and notochordal mesoderm, as well as some lateral mesoderm, invaginate around the dorsal blastoporal lip. The lateral lips also form part of the ventral mesoderm, while the ventral lip forms all the posterior and lateral extra-embryonic mesoderm (Hubert, 1962). From the blastopore a narrow, meso-dermal archenteron, more or less triangular in cross section, extends forward in between the ectodermal and endodermal layers (fig. 2.24a–c). Later the archenteron expands laterally, its lateral walls moving upward and forming the lateral portions of the definitive archenteron roof. Numerous cells leave the floor of the archenteron posteriorly and laterally, forming extra-embryonic mesoderm. The anterior end of the mesodermal archenteron fuses with the underlying hypoblast, thus forming the endo-mesodermal prechordal plate. As a result of the lateral migration of cells from the floor of the archenteron the latter becomes perforated anteriorly. Even prior to this the hypoblast or 'vitelline leaflet' seems to break through (Hubert, 1962). As a consequence of these events the anterior archenteron opens directly into the subgerminal cavity. An outcome of the lateral and caudal extension of this opening is that a large triangular perforation of the archenteric floor develops (fig. 2.24d). The primary hypoblast is displaced laterally and anteriorly, so that an anterior endoblastic ridge arises in front of the prechordal plate (Pasteels, 1957b) (fig. 2.24d).

When the archenteron has reached its ultimate forward and lateral extension a neural plate is formed over the prechordal and chordomesodermal archenteron roof. The axial organs subsequently lengthen by a further stretching and dorsal convergence of neural and mesodermal anlagen, while material on either side of the blastopore continues to invaginate until the neural plate stage (late invagination, Pasteels, 1957b).

Underneath the mesodermal archenteron roof a single-layered endodermal sheet is finally formed: the definitive embryonic endoderm. Opinions again differ as to its mode of formation. According to Hubert (1962) it is formed by forward movement of endodermal material from the ventral blastoporal lip region, but according to Pasteels (1957b) the fragmented primary hypoblast situated underneath the lateral somitic mesoderm moves medially and fuses again underneath the notochord.

Thus uncertainty still exists both about the formation of the primary hypoblast and about the origin of the definitive embryonic endoderm. Nothing is known about the possible formation of a secondary or tertiary hypoblast as observed in the avian embryo.

In the extra-embryonic region in front of the embryonic anlage an anterior amnion fold develops, which extends laterally and caudally, overgrowing the embryo from front to back.

Summarising, it may be said that the gastrulation process in the reptiles resembles that in the lower vertebrates, but that both the early formation of a hypoblast and the formation of an amnion are distinct parallels of development in birds and mammals. Unfortunately, nothing is known about the causal events in early development which lead to mesoderm and definitive endoderm formation.

General conclusions

Surveying the descriptive and experimental data on the origin of the mesoderm in the early development of the various subphyla, classes and orders of the chordates, the following general hypothesis may be proposed.

In all chordates the formation of the mesoderm is based on a common mechanism: it arises from the still-totipotent animal or epiblastic moiety of the embryo under an inductive action exerted by the already determined vegetal or hypoblastic moiety. Through this interaction a polarity is transferred from the former to the latter, which is a prerequisite for the spatial organisation of the mesodermal mantle or layer. In addition to the fairly convincing evidence for this hypothesis in the amphibians and birds, there is circumstantial evidence in a number of other groups: the development of the cephalochordates greatly resembles the situation in the amphibians and the agnathan and osteichthyan fishes, while that of the mammals resembles the situation in the birds. Evidence is scanty in the chondrichthyan and teleostean fishes and the reptiles, and almost non-existent in the tunicates. However, the descriptive and experimental evidence available for these groups does not contradict the hypothesis as formulated above.

3

Characteristics of the primordial germ cells

General cytological features

At the time when only cytological staining techniques were available the PGCs were described as almost identical in the various groups of the chordates. The PGCs are large compared with the surrounding cells, are roundish, oval or pear-shaped, and have a distinct cell boundary and a diameter varying from 10 to 20 μm depending on shape, stage and species. The voluminous nucleus is roundish, oval or lobular in shape and has a distinct nuclear membrane. It is often situated eccentrically; its diameter may vary between 6 and 10 μm. The chromatin is fairly evenly distributed throughout the nuclear plasm, giving the stained nucleus a light or transparent appearance. The nucleus contains one or two prominent nucleoli. Depending on the stage of development and the taxonomic group the cytoplasm contains varying amounts of yolk granules, fat droplets and pigment granules (fig. 3.1). In general the cytoplasm has a dense, granular appearance and contains a roughly hemispherical cytoplasmic body called variously the centre sphere (Johnston, 1951), cytoplasmic crescent (Simon, 1964), mitochondrial crescent (Celestino da Costa, 1937) or attraction sphere (Risley, 1933). This consists of one or two centrioles, numerous mitochondria, and small yolk platelets. The unequal distribution of yolk and lipid inclusions and the position of the cytoplasmic body result in the eccentric location of the nucleus and the oval or pear-shaped form of the cell. The PGCs are often also characterised by the presence of lobopodia or filopodia, particularly during their migratory period. The general features of the PGCs have been described by, among others, Johnston (1951) for teleost fishes, Bounoure (1939) for anuran amphibians, Humphrey (1925) for urodele amphibians, Marcus (1938) for coecilians, Risley (1933) for chelonian reptiles, Simon (1962) for birds, and Celestino da Costa (1937) and Witschi (1948) for mammals.

It is clear that none of the features of PGCs described above is specific for these cells, so that an unequivocal identification can only be made on the basis of a combination of all or most of them. Moreover, the earlier the stage of development, the more yolk the cells usually contain and the less pronounced are the various cell characteristics. A large amount of yolk also often leads to 'deformation' of both cell and nucleus, and a consequent blurring of

54

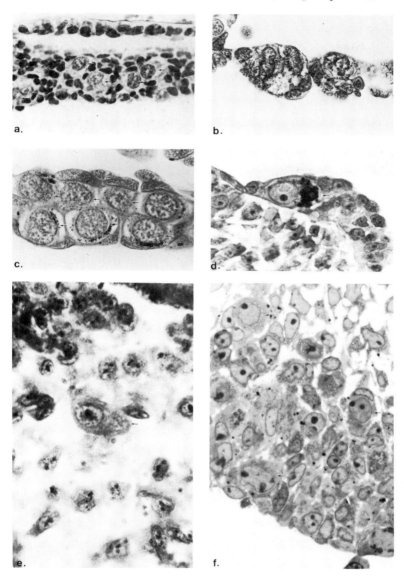

Fig. 3.1. Light micrographs of PGCs (short arrows) in the genital ridges of (a) the fish, *Raja batis* (× 350); (b) the anuran, *Discoglossus pictus* (× 350); (c) the urodele, *Ambystoma mexicanum* (× 350); (d) the reptile, *Lacerta vivipara* (× 600) (courtesy Dr J. Hubert); (e) the bird, *Gallus domesticus* (× 1050) (courtesy Dr G. Reynaud); (f) the mammal, *Mus musculus* (× 700) (courtesy Dr E. M. Eddy)

several of the above features. The early workers were therefore unable to trace the PGCs back to the earliest stages of development.

The discovery by Bounoure in anurans of a special cytoplasmic structure, the so-called 'germinal plasm', which is originally located in the subcortical layer of the vegetal region of the fertilised egg and later in the PGCs of the tadpole (see Bounoure, 1939), marked an important step forward in the analysis of the origin of the PGCs in the anuran amphibians. Cytological and cytochemical studies indicated that the germinal plasm contained, besides small yolk platelets and pigment granules, mitochondria and RNA. However, it was the electron microscope which revealed the true nature of the germinal plasm. In recent years numerous investigations have been devoted to the study of this structure that is specific to germ cells. It must however be emphasised that not all chordate or vertebrate PGCs are characterised by the presence of a typical germinal plasm, and that moreover it is not found during all developmental stages.

Since nothing is known about the ultrastructure of the PGCs in the Cephalochordata and Tunicata the following description will only deal with the ultrastructural and cytochemical characteristics of the PGCs in the various groups of the vertebrates. For extensive surveys of the literature on germinal plasm in invertebrates and vertebrates, the reader is referred to the review articles by Beams & Kessel (1974), Eddy (1975) and Smith & Williams (1975).

As in the preceding chapter the normal taxonomic order cannot be followed, since our knowledge of the various groups is too variable; in some groups it is quite extensive, whereas in other groups it is scanty or even non-existent. The situation in the anuran amphibians will therefore be discussed first. The other amphibians will then be treated, followed by the fishes. The discussion of the amniotes will begin with the birds, continue with the mammals, and close with the reptiles.

The ultrastructural and cytochemical characteristics of the PGCs in Anamnia

Amphibia

In the amphibians extensive ultrastructural and cytochemical analyses have been carried out on the PGCs of the anurans, few on those of the urodeles, and none on those of the coecilians.

Anura

In the anurans electron-microscopic analysis has revealed, apart from the usual cell organelles found in almost every cell of the embryo, a number of special cytoplasmic and nuclear structures which may be considered as being characteristic of PGCs and their descendants. The first and most striking of these is the 'germinal plasm', which has now been described in different anuran species by many investigators, for example Balinsky (1966), Cambar, Delbos & Gipouloux (1970), Kessel (1971), Mahowald & Hennen (1971), Williams & Smith (1971), Czolowska (1972), Kalt (1973), Beams & Kessel (1974), Gipouloux (1975) and Ikenishi & Kotani (1975).

The germinal plasm is a well-circumscribed structure occurring in the form of patches 5–12 μm in diameter. These consist of numerous mitochondria, many containing protein crystals (Kessel, 1971), interspersed with so-called 'germinal granules' (electron-dense bodies). These electron-dense bodies are composed of small electron-dense foci which seem to be embedded in a matrix of extremely fine fibrils. It is not clear whether the structure of the germinal granules is essentially granular, fibrillar or fibrillo-granular. The electron-dense bodies are not bounded by a membrane and are frequently found in contact with mitochondrial membranes ('intermitochondrial cement') or with the nuclear membrane ('nuage material'). The germinal plasm also contains pigment granules and some glycogen, as well as a few lipid droplets and a varying amount of small yolk platelets.

Blackler (1958) thought that the islands of germinal plasm found in cleaving eggs arise only after fertilisation, but Czolowska (1969) demonstrated the presence of small patches of germinal plasm in ovarian oocytes. Williams & Smith (1971) found small electron-dense areas devoid of ribosomes within clusters of mitochondria just beneath the vegetal pole cortex of ovarian oocytes. Around the time of germinal vesicle breakdown larger electron-dense areas are seen which are now associated with ribosomes. Since typical germinal granules also appear in progesterone-treated oocytes which are subsequently enucleated, this 'condensation' process does not depend upon germinal vesicle contributions. The characteristic germinal granules therefore seem to coalesce from precursors during oogenesis with a possible synthesis of components *de novo* during maturation (see also Smith, 1975).

According to Mahowald & Hennen (1971) the precursors of the germinal plasm in *Rana pipiens*, described as cytoplasmic islands in the unfertilised and fertilised egg and in early cleavage stages, consist essentially of the same components as the germinal plasm in the PGCs of the tadpole. However, a certain structural evolution seems to occur in the germinal plasm during development. In the fertilised egg the germinal plasm is present in the form of small cytoplasmic patches with germinal granules 0.2–0.3 μm in diameter (fig. 3.2). This is in agreement with Czolowska's (1969) observations on *Xenopus*, where a large number of very small patches was found in the cortical and subcortical cytoplasm of the vegetal region of ovarian oocytes. In the unfertilised egg the electron-dense foci of the germinal granules measure about 20 nm. At the late blastula stage the germinal granules have acquired a diameter of 0.4–0.5 μm, while the electron-dense foci now have a diameter

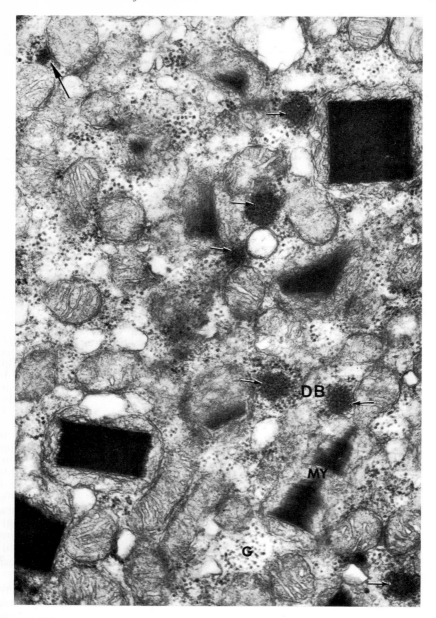

Fig. 3.2. Ultrastructure of the germinal plasm in the anuran, *Rana pipiens*; mitochondria (M) with yolk crystals (Y), electron-dense bodies (DB, short arrows), partially as intermitochondrial cement (long arrow), and some membranous elements and glycogen (G) in the cytoplasmic matrix ($\times$ 33 000). (Courtesy Dr A. P. Mahowald & Dr S. Hennen.)

of up to 50 nm. A condensation process has apparently occurred. In the unfertilised and fertilised egg and at early cleavage stages the germinal plasm seems to contain some RNA in the form of clusters of ribosomes associated with the germinal granules. The number of ribosomes increases upon fertilisation. At the early tadpole stage the germinal granules are in the form of amorphous bodies of a clearly fibrous nature. They are situated in close proximity to the nuclear membrane or are attached to mitochondria, but are no longer associated with clusters of ribosomes. Mahowald & Hennen (1971) ascribe a morphogenetic significance to the temporary association of clusters of ribosomes with the germinal granules, mainly in analogy to similar events observed in the development of the polar granules in the pole cells of certain insects. Protein synthesis has been demonstrated in the metabolically active germinal plasm of the *Xenopus* egg during cleavage (Hogarth & Dixon, 1976).

According to Ikenishi & Kotani (1975) the fibrillo-granular germinal granules found in *Xenopus* at early stages of development first change into irregular string-like bodies at the late neurula to early tail-bud stages, and then into granular material at the feeding tadpole stage (fig. 3.3).

In *Xenopus* Kalt (1973) distinguishes three different structures which are found in both the male and the female germ line but in no other cells. The first of these is the germinal plasm, which takes the form of large proteinaceous masses of electron-dense material, which he calls 'nuage material'. The second structure, also cytoplasmic in nature, is the electron-dense so-called chromatoid or nucleolus-like body. This is found in germ cells of the indifferent gonad and throughout the female germ line, as well as in the early stages of the male germ line, where it dissociates into coated vesicles during spermatogenesis. The third structure is a fibrillar nuclear region (fibrillar meshwork) which is free of DNA (fig. 3.4). A similar structure has been described in *Xenopus* by Reed & Stanley (1972) and by Coggins (1973).

Opinions differ on the homology of the various electron-dense structures found in the cytoplasm of PGCs. Some authors, for example Kalt (1973) and Kalt, Pinney & Graves (1975), make a distinction between chromatoid or nucleolus-like bodies on the one hand and germinal granules or nuage material on the other, on the basis of differential sensitivity to various cytochemical treatments. Kalt (1973), Eddy (1974) and Webb (1976) emphasise the great similarity in ultrastructure between germinal granules and nuage material, as well as their resemblance to nucleolar components in the nucleus. Other authors, for example Fawcett (1972), consider the various electron-dense components of the cytoplasm of germ cells as belonging to one and the same family. It must however be emphasised that neither is the homology of the various structures established nor is their functional significance sufficiently understood (see also reviews by Beams & Kessel, 1974; Eddy, 1975; Smith & Williams, 1975).

The germinal plasm originally present in the fertilised egg does not change

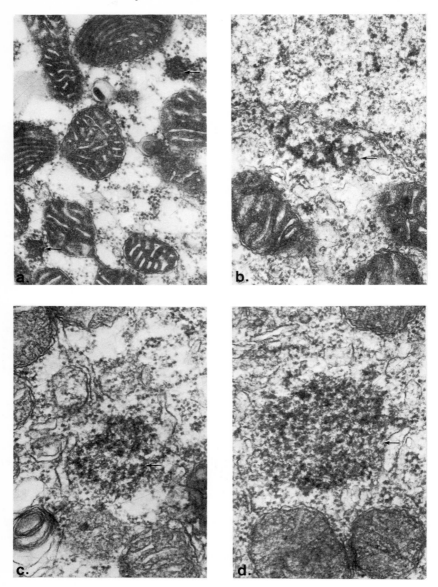

Fig. 3.3. Changes in the ultrastructure of the electron-dense bodies of the germinal plasm (arrowed) in PGCs of *Xenopus laevis*. (a) Fibrillo-granular germinal granules at stage 12 (Nieuwkoop & Faber, 1975) ($\times 37\,500$). (b) Irregularly shaped string-like body (ISB) at stage 46 and earlier ($\times 52\,500$). (c) Transition between ISB and granular material at stage 46 ($\times 45\,000$). (d) Granular material at stage 46 and later ($\times 45\,000$). (Courtesy Dr K. Ikenishi & Dr M. Kotani.)

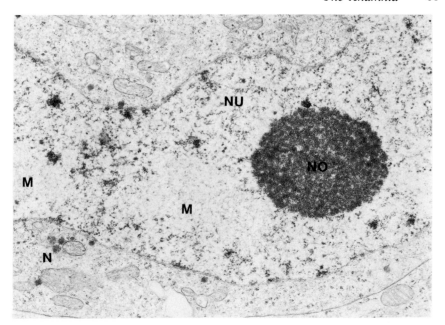

Fig. 3.4. Ultrastructural features of the PGCs in *Xenopus laevis* at stage 42 (Nieuwkoop & Faber, 1975). Electron micrograph of a PGC in the genital ridge, with fibrillo-granular nucleolus (NO) and fine fibrillar meshwork (M) in the nucleus (NU), and small patches of nuage material (N) in the cytoplasm ($\times 24000$). (Courtesy Dr M. R. Kalt.)

in amount during cleavage. It is distributed unevenly over the exponentially increasing number of blastomeres (Whitington & Dixon, 1975). Although the number of blastomeres containing germinal plasm initially increases, their total number later remains constant as a consequence of the transfer of the germinal plasm to one daughter cell of a division only (pPGCs). During gastrulation and neurulation a small number of clonal divisions occur, as a result of which the number of PGCs increases about fourfold; these are followed by one more clonal division before stage 35 (Dziadek & Dixon, 1975, 1977). It is evident that during these clonal divisions the amount of germinal plasm must increase, since otherwise dilution would occur. The frequently observed passage of electron-dense material through nuclear pores may be associated with the formation of new germinal plasm. However, the static electron micrographs do not allow a distinction between a transfer of material from the nucleus to the cytoplasm or vice versa. During the clonal divisions the PGCs also synthesise RNA (Dziadek & Dixon, 1977).

It seems of importance also to mention some negative features of the PGCs in anurans. Gipouloux (1967) was unable to demonstrate either glycogen or alkaline phosphatase in anuran PGCs. The former is at variance with Kessel's (1971) electron-microscopic observations, but the latter is in accordance with Chiquoine & Rothenberg's (1957) studies on PGCs in urodele amphibians.

The distinct cell boundary of PGCs already described by the earlier workers in the various groups of vertebrates turns out to be due to the presence of large intercellular spaces between the PGCs and the surrounding somatic cells. This emphasises the non-participation of the PGCs in embryonic tissues (Ikenishi & Kotani, 1975; Ikenishi & Nieuwkoop, 1978) (fig. 3.5a). The PGCs show pseudopodia and lobopodia which protrude into the intercellular spaces. The demonstration of microtubules and microfilaments in the cytoplasm of spermatogonia constitutes additional evidence for their migratory activity (Kalt, 1973).

Urodela

In the urodele amphibians only very scanty and partially conflicting data on the ultrastructure of the PGCs are available.

Recent electron-microscopic observations of Ikenishi & Nieuwkoop (1978) on the ultrastructure of the PGCs in *Ambystoma mexicanum* reveal that the PGCs of the urodele larva at stage 46 (Harrison, 1969) are characterised by the presence of germinal plasm consisting of clusters of mitochondria and electron-dense bodies of a fibrillo-granular nature. However, the germinal plasm only appears at early larval stages (from stage 40 onwards), at which time the passage of electron-dense material through nuclear pores is also observed. The amount of electron-dense material increases rapidly during subsequent developmental stages, until the definitive complement is reached by stage 46 (fig. 3.5).

The late appearance of the germinal plasm in the urodeles seems to be at variance with Williams & Smith's (1971) brief and unconfirmed statement on the presence of germinal granules in the subcortical cytoplasm of the equatorial region of the fertilised, uncleaved egg of *Ambystoma mexicanum*.

Intra-nuclear fibrillar regions, as found by Kalt (1973) in anuran PGCs, have recently also been observed by Ikenishi & Nieuwkoop (1978) in *Ambystoma mexicanum* PGCs (fig. 3.5a), but nothing is known about their functional significance.

Fishes

Little is known about the ultrastructure of the PGCs in fishes and the relevant literature is virtually restricted to the Teleostomi. The PGCs in the Agnatha, Chondrichthyes and Osteichthyes have so far only been studied with the light microscope, as described in the first part of this chapter.

Gamo (1961b) emphasised the fact that the PGCs in meroblastic embryos

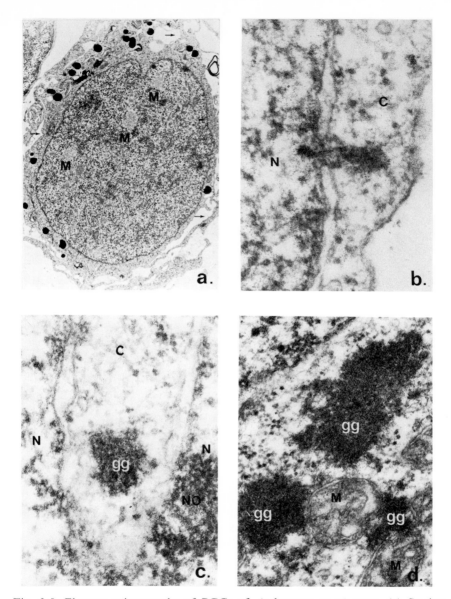

Fig. 3.5. Electron micrographs of PGCs of *Ambystoma mexicanum*. (a) Section through an entire PGC in the genital ridge at stage 46 (Harrison, 1969), showing lobular nucleus with fibrillar meshwork (M), cytoplasm with mitochondria and yolk, and large intercellular spaces (short arrows) between the PGC and surrounding peritoneal cells (× 4125). (b) Transfer of electron-dense material between nucleus (N) and cytoplasm (C) at stage 40 (× 90000). (c) First appearance of fibrillo-granular material in the cytoplasm of a PGC of stage 40; cytoplasmic area (C) enclosed by lobular nucleus (N) with nucleolus (NO) (× 90000). (d) Fibrillo-granular germinal granules (gg) in close association with mitochondria (M) in germ plasm of a stage 46 PGC (× 56250). (From Ikenishi & Nieuwkoop, 1978.)

contain less yolk than those in holoblastic embryos. This may explain their tendency to associate themselves with nutritional sources, e.g. with the periblast in teleost embryos.

Pala (1970) mentions the presence of small PAS-positive granules in the perinuclear cytoplasm of teleost PGCs, particularly in the 'attraction sphere'. Nedelea & Steopoe (1970) were the first to describe the presence in the cytoplasm of PGCs of electron-dense bodies, which they call 'nuage material'. Satoh (1974) gives a more detailed description of the electron-dense bodies. They are up to 1 μm in diameter, not bounded by a membrane, and consist of an interwoven meshwork of very fine fibrils. They are often found in close association with clusters of mitochondria, in direct contact with their outer membranes. This description fits that of germinal plasm very well. Nothing is known, however, of the time of appearance of these structures in the development of the teleost PGCs.

The ultrastructural and cytochemical features of PGCs in Amniota

Aves

So far electron-microscopic studies have only revealed the presence of common cell organelles (de Simone-Santoro, 1967). No mention has been made of structures specific to germ cells, such as germinal plasm or nuage material. However, some cytochemical features can be used to characterise the PGCs in birds.

Although alkaline phosphatase activity had been demonstrated in chick PGCs by Chiquoine & Rothenberg (1957), the activity is too low to characterise them properly, since a similar activity was found in other cells. The most prominent feature is their high content of PAS-positive material (fig. 3.6). This was first demonstrated by McKay, Hertig, Adams & Danziger (1953) and later confirmed by many authors, e.g. Meyer (1960), Simon (1964), Clawson & Domm (1963a, b), Reynaud (1968), Dubois & Cuminge (1968), Dubois & Croisille (1970) and Fujimoto et al. (1975, 1976a, b). According to McCallion & Wong (1956) it is not an absolute criterion, since PAS-positive material is also found in other cells. We may say, however, that the combination of the various cytological and cytochemical features adequately characterises the PGCs in birds.

The PGCs in birds also go through some form of evolution during development. Whereas in early stages PGCs contain a large amount of yolk and lipids and only a small amount of glycogen, in later stages they contain less yolk and lipids, but abundant glycogen (Clawson & Domm, 1963b; Fujimoto et al., 1975, 1976a, b). The amount of glycogen drops again during migration. Clawson & Domm (1963a, b) and Fujimoto et al. (1976a) also observed changes in the ultrastructure and regional distribution of the glycogen: it is non-granular and evenly distributed in early stages, but granular and concentrated at one pole of the cell in later stages.

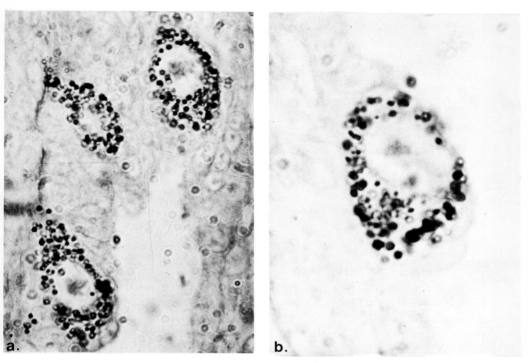

Fig. 3.6. PGCs in the gonadal epithelium of the chick embryo, with glycogen inclusions in the form of granules stained with Gendre-PAS (black dots and rings). (a) × 2550, (b) × 5100. (Courtesy Dr G. Reynaud.)

Mammalia

In the mammals McKay *et al.* (1953) were the first to demonstrate high alkaline phosphatase activity in the cytoplasmic rim of PGCs (fig. 3.7). This was confirmed by Chiquoine (1954), Mulnard (1955), McAlpine (1955), Mintz (1957*b*), Asayama (1963) and Chrétien (1968). Jeon & Kennedy (1973) also observed high alkaline phosphatase activity in the Golgi complex. Clark & Eddy (1975) and Fujimoto *et al.* (1977) mention that the alkaline phosphatase activity along the cell periphery is only faint in early stages but increases markedly in later stages. Asayama (1963) emphasises the fact that high alkaline phosphatase activity is not a feature specific to germ cells, since other cells may also show it. It is moreover stage dependent. During germ cell migration as well as during gametogenesis the germ cells seem to associate themselves with other tissues due to a lack of adequate endogenous energy sources. The alkaline phosphatase activity may thus be connected with an active intercellular exchange of material (Semenova-Tian-Shanskaya, 1969; Falin, 1969; Merchant & Zamboni, 1973; Zamboni & Merchant, 1973).

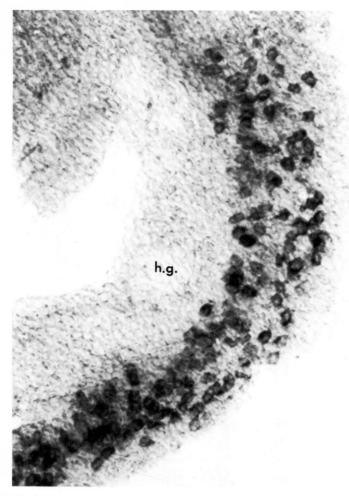

Fig. 3.7. Hindgut preparation of day 9 mouse embryo, reacted for alkaline phosphatase. Positively reacting PGCs are for the most part within the gut epithelium (h.g.). (× 400) (Courtesy Dr J. M. Clark & Dr E. M. Eddy.)

Asayama (1965) observed the presence of rather large amounts of PAS-positive material in mammalian PGCs. This was confirmed by Falin (1969), Wartenberg, Holstein & Vossmeyer (1971), Fuyuta, Miyayama & Fujimoto (1974), Fukuda (1976) and Fujimoto *et al.* (1977). This feature is not specific either, since other cells also contain much glycogen. It disappears, moreover, with the onset of meoisis in the development of the female gametes, and at a later stage in spermatogenesis. Fujimoto *et al.* also observed a large number of lipid droplets in human PGCs.

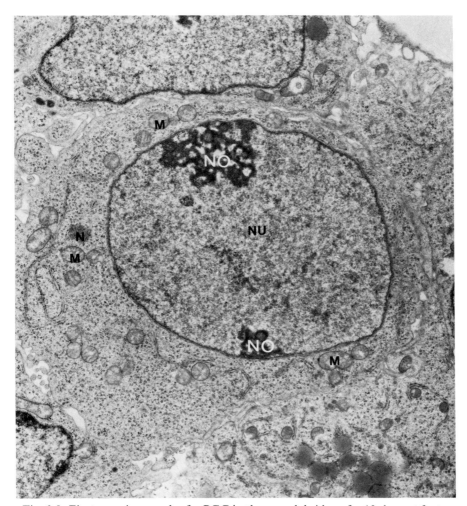

Fig. 3.8. Electron micrograph of a PGC in the gonadal ridge of a 13-day rat foetus, showing nucleus (NU) with large nucleoli (NO) as well as nuage material (N) and mitochondria (M) in the cytoplasm. (× 7500) (Courtesy Dr E. M. Eddy.)

Eddy (1974) described in detail the presence of electron-dense, granulo-fibrous material, called 'nuage material', in the cytoplasm of mammalian primordial germ cells (fig. 3.8), which is generally poor in inclusions. It was found either in the form of free, discrete bodies or as intermitochondrial cement, and was sometimes spatially associated with pores in the nuclear membrane. Eddy found these structures in young PGCs but also in the PGCs of the indifferent gonad and in the germ cells of neonatal and adult animals. These observations were corroborated by Motta & Van Blerkom (1974), who found nucleolus-like bodies in the cytoplasm of large oocytes as well as in PGCs of early developmental stages. During spermatogenesis Comings & Okada (1972) had found similar structures, probably formed by the extrusion of nucleolar components into the cytoplasm, which they called chromatoid bodies. Fawcett, Eddy & Phillips (1970), who had seen similar structures, did not assume them to be of nuclear origin. It is not yet clear whether these structures really are a constant feature of mammalian germ cells. While Eddy points out the strong resemblance of the nuage material to the germinal plasm of the anuran egg. Sud (1961) and Motta & Van Blerkom (1974) moreover call attention to its similarity to the chromatoid body found during sperm-atogenesis. However, in the latter RNA could not be demonstrated (Eddy, 1970).

It may finally be mentioned that during their migratory period mammalian PGCs show cytoplasmic lobopodia and filopodia containing numerous microfilaments (see Jeon & Kennedy, 1973; Merchant & Zamboni, 1973; Zamboni & Merchant, 1973; Spiegelman & Bennett, 1973; Clark & Eddy, 1975; Fukuda, 1976; Fujimoto *et al.* 1977).

Reptilia

Milaire (1957) mentioned that in the Chelonia both the extra-gonadal and the gonadal PGCs contain large amounts of PAS-positive material. The PGCs of Lacertilia, besides some glycogen, also contain numerous yolk granules and lipid droplets (Hubert, 1970*a*).

The ultrastructure of the PGCs has been fairly extensively studied only in the Squamata. Hubert (1970*b*) describes the ultrastructural organisation of the PGCs in the Lacertilia. Structures specific to germ cells are not found in the cytoplasm but only in the nucleus. He mentions the presence of a paranucleolar mass as well as of a small, dense fibrillar corpuscle ('ring-shaped' in cross section); both are situated in the vicinity of the large 'ring-shaped' nucleolus (fig. 3.9). When there are two nucleoli both structures are also present in duplicate. The paranuclear mass consists of a network of fine fibrils and very fine particles. It is probably of a proteinaceous nature, since it is Feulgen-negative but Fast Green-positive and stains green with Methyl Green Pyronine. A similar structure was described as 'fibrillar meshwork' in PGCs of anuran (Kalt, 1973) and urodele amphibians (Ikenishi & Nieuwkoop, 1978) (see pp. 59, 62). No germinal plasm or nuage material has hitherto been demonstrated. In *Lacerta* the paranuclear mass is a constant feature of PGCs, of oogonia and oocytes, and of spermatogonia (Hubert, 1976). Hubert (1968) found the same structure in the snake *Vipera* but not in the slow-worm *Anguis*, so that it may actually not be a constant feature in the Squamata. The small fibrillar corpuscle is also found in spermatogonia of *Lacerta* (Hubert, 1970*b*).

During their migratory period the PGCs show micropseudopodia con-

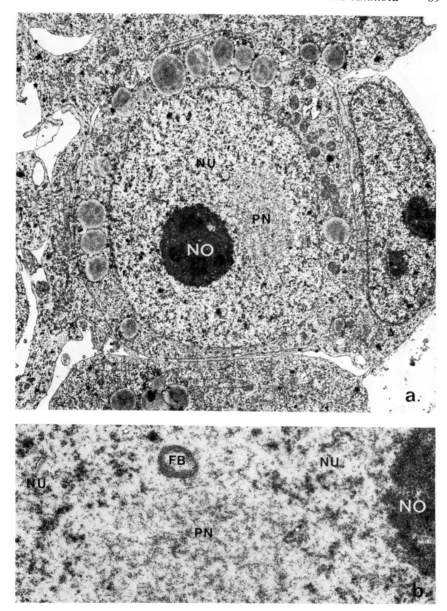

Fig. 3.9. (a) Nucleolus (NO) and large paranucleolar mass (PN). (b) Small 'ring-shaped' dense fibrillar corpuscle (FB) inside the nucleus (NU) of a PGC of *Lacerta vivipara*. (a) × 10 125, (b) × 52 500. (Courtesy Dr J. Hubert.)

taining bundles of microfibrils, which provide evidence of amoeboid motility (Hubert, 1970*b*).

General conclusions

Summarising, it may be stated that the *general cytological features* of the PGCs (see p. 54) are very much alike in the various groups of chordates. Although any of these features in itself is not specific for germ cells, the combination of various features quite satisfactorily serves for the unequivocal identification of PGCs. It must however by emphasised that during certain periods of development, particularly in the very early stages and during rapid multiplication, several of these features are rather indistinct or even unrecognisable. This has led to the suggestion of a discontinuity in the germ line and a formation of germ cells *de novo* during certain periods of development, particularly during gametogenesis.

Ultrastructural analysis, which unfortunately is still rudimentary or even lacking in certain groups, suggests quite strongly that the electron-dense fibrillo-granular bodies or the intermitochondrial cement found in the germinal plasm are structures specific to germ cells. They have not been found in all groups, however, so that they certainly do not represent a universal feature of germ cells. We shall return in chapter 5 (p. 87) to the very important question of how far the so-called germinal plasm may actually be called a germ cell determinant.

It is intriguing that the PGCs, which show almost identical general cytological characteristics in all chordate groups, nevertheless show pronounced differences in ultrastructure or cytochemistry. We have already mentioned the non-universality of the germinal plasm. Glycogen, which is so abundant in avian germ cells, is also found in mammalian and reptilian germ cells, but in smaller amounts, while it is virtually absent from the PGCs of amphibians and fishes. A high alkaline phosphatase activity of the cell periphery seems to characterise the mammalian PGCs. A special nuclear structure in the form of a fibrillar meshwork inside the nucleus, which is free of DNA but RNA-positive, is found in the PGCs of anuran and urodele amphibians as well as in those of certain reptiles.

During the last decades attention has been focussed particularly on the cytoplasmic features of the PGCs, so that specific nuclear structures may have been overlooked in a number of vertebrate groups. In our opinion it is not unlikely that nuclear structures may represent more constant and more specific features of the PGCs than the well-known germinal plasm.

4

The extra-gonadal and one-time origin of the primordial germ cells

Two controversial issues with respect to the origin of the PGCs in vertebrates have dominated the literature for many decades: (*a*) that of the gonadal versus extra-gonadal origin of the PGCs, and (*b*) that of the one-time versus repeated origin of the PGCs during the life span of the organism. We feel that these controversial issues should be dealt with before the place and mode of origin of the PGCs can be satisfactorily discussed.

Gonadal versus extra-gonadal origin of the PGCs

The early literature on the origin of the PGCs in vertebrates was dominated by the controversial issue of the gonadal versus extra-gonadal origin of the PGCs. Embryologists were sharply divided into opposing camps, i.e. those supporting Waldeyer's (1870) ideas and those advocating Nussbaum's (1880) concepts. Waldeyer's theory, which implies an origin of the PGCs from the somatic 'germinal epithelium' of the gonadal anlage, was primarily based on a study of the higher vertebrates, where the PGCs were not clearly distinguishable before they were inside the gonadal anlagen. The concept was later extended to all the vertebrates and even to a number of invertebrate groups. Nussbaum's notion, which implies an early segregation of the PGCs from the somatic cells of the embryo long before the formation of the gonadal anlagen and topographically separated from them, was chiefly based on studies in the anuran amphibians and teleosts.

The issue remained controversial until the fourth decade of this century, as a result of the rather unspecific staining methods available. During the nineteen-thirties the balance between the two points of view began to shift towards extra-gonadal origin, particularly in the lower vertebrates. When Bounoure's book appeared in 1939 the advocates of Waldeyer's concept had lost nearly all their ground as regards the lower vertebrates, but not as regards the higher vertebrates, especially the mammals, where the PGCs could only be clearly distinguished inside the gonadal anlagen.

Brambell (1960) already stated: 'It is now recognized by a large majority of embryologists that [in the vertebrates] the PGCs arise exceedingly early [in development], long before the rudiments of the gonad are formed and at a distance from the site they will [ultimately] occupy... Their state of origin is

71

extra-gonadal in all instances and in the amniotes and some of the fishes it is [moreover] extra-embryonic.'

This marked shift in the general consensus was mainly due to the development of new histological and histochemical techniques which allowed identification of the PGCs at much earlier stages of development than was possible until then. Recent studies have fully confirmed Brambell's conclusions, e.g. in the fishes by Johnston (1951), Gamo (1961*b*), De Smet (1970), Nedelea & Steopoe (1970) and Pala (1970); in the amphibians by Lacroix & Capuron (1966), Gipouloux (1967) and Sutasurya & Nieuwkoop (1974); in the reptiles by Pasteels (1953, 1957*a*, 1964) and Hubert (1969, 1976); in the birds by Simon (1960, 1964), Dubois (1965), Rogulska (1968), Fargeix (1969, 1970, 1975), Reynaud (1969), Clawson & Domm (1969), Komar (1969), Dubois & Croisille (1970), Bruel (1973) and Fujimoto *et al.* (1976*b*); and finally in the mammals by Mintz & Russell (1957), Ozdzenski (1967, 1969), Falin (1969), Semenova-Tian-Shanskaya (1969), Merchant & Zamboni (1973) and Zamboni & Merchant (1973).

One-time versus repeated origin of the germ cells

In the last part of the nineteenth and the first half of the twentieth century the concept of the origin of the PGCs was obscured by the controversial issue of whether there exists only a single source or multiple sources of germ cells during the life span of an individual. The concept of the continuity of the germ line as postulated in Weismann's *Keimplasma* theory implies a segregation of the germ cells from the somatic cells – preferably during early development – as well as the persistence of the former throughout fertile life. The adversaries of this concept essentially claimed the existence of several phases of asexual and sexual life.

In his 1945 review Everett distinguished four different categories among those working on the origin of the germ cells in vertebrates: (1) those who deny the early segregation of the PGCs and advocate only a late, somatic origin from the 'germinal epithelium', (2) those who believe in the early segregation of germ cells but postulate the subsequent degeneration of this first generation and the ultimate formation of a second generation of definitive germ cells from the 'germinal epithelium', (3) those who agree to the early segregation of PGCs and their subsequent development into definitive gametes, but in addition postulate a supplementary formation of germ cells from the 'germinal epithelium', and (4) those who defend a one-time, early segregation of the PGCs, which constitute the only source of the definitive gametes. The issue is further complicated by the assumption either of a periodic or of a continuous formation of germ cells from the 'germinal epithelium' throughout the reproductive period.

We have seen in the preceding section that in the vertebrates the PGCs are

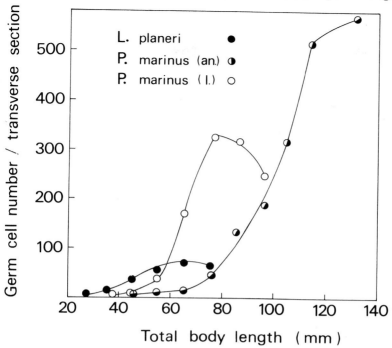

Fig. 4.1. Increase in average germ cell number per transverse body section in relation to total body length, in *Lampetra planeri* and in *Petromyzon marinus* of the anadromous (an.) and landlocked (l.) races (After Hardisty, 1971.)

formed outside the gonadal anlagen, to which they are subsequently transferred. The problem therefore boils down to the question of whether these PGCs represent the sole source of the definitive gametes or whether they are supplemented with or replaced by one or more generations of germ cells formed *de novo* in the gonadal anlagen during later phases. This uncertainty was mainly due to the fact that particularly in the higher vertebrates the PGCs lose most of their cytological characteristics during multiplication in the gonadal anlagen. The problem can therefore only be definitely solved by experimental analysis. Nevertheless, the evidence for the existence of a single and uninterrupted germ line in the life cycle of vertebrates is both descriptive and experimental. The descriptive evidence consists of quantitative data on germ cell history in normal animals and sterile mutants, while the experimental evidence mainly relates to defect and grafting experiments.

Quantitative data

In the vertebrates *oogenesis* follows a uniform pattern with two main variants. Either germ cell multiplication continues uninterruptedly or cyclically throughout the reproductive period (as in most teleosts, all amphibians, the majority of reptiles and possibly some mammals), or it ceases early in life with the onset of meoisis, so that a definitive, finite stock of germ cells is formed for the entire fertile period (as in cyclostomes, elasmobranchs, a few teleosts, perhaps a few reptiles, all birds, the monotremes, and all eutherian mammals, with a few possible exceptions) (Franchi, Mandl & Zuckerman, 1962).

Following Okkelberg's (1921) study on the early history of the germ cells in the brook lamprey, Hardisty & Cosh (1966) followed the germ cell history in several Cyclostomata from the appearance of the germ cells to the establishment of the definitive stock of oocytes in the sexually differentiated gonad. The final maturation and spawning of the eggs occurs only several years later, i.e. shortly before the end of adult life. When plotting the number of germ cells against the total length of the larva Hardisty (1971) found smooth S-shaped curves (fig. 4.1). He concluded from these data and from the frequent observation of mitotic figures that there is actually no need to postulate a formation of PGCs *de novo* from somatic cells. Their initial increase and later relative decrease in number can easily be accounted for by the observed mitotic proliferation and subsequent degeneration of germ cells. The observed differences in fecundity of the various species of lamprey must be due to differences in proliferation, since the initial average number of PGCs is nearly the same in the different species. The difference between the actual fecundity of the adult female (number of eggs produced) and the potential fecundity of the ammocoete larva (maximum number of germ cells) can be accounted for by the observed extensive degeneration of germ cells during sexual differentiation.

In the chick Hughes (1963) studied germ cell number in the left ovary from day 9 of incubation till the first day after hatching. The formation of oocytes ceases at about the time of hatching. Oogonial mitoses are responsible for the nearly 25-fold increase in germ cell number between day 9 and day 17, while the subsequent decrease by about 40% between day 17 and the first day after hatching is due to the high incidence of degenerating germ cells. Hughes concludes that also in the chick ovary there is no need to postulate a second source of germ cells during oogenesis.

A similar study, on oogenesis in the rat by Beaumont & Mandl (1962), performed between day $14\frac{1}{2}$ *post coitum* (*p.c.*) and day 2 *post partum* (*p.p.*), shows that oogonia are mitotically active until day $17\frac{1}{2}$ *p.c.*, after which a sharp decline in mitotic activity occurs at the onset of the leptotene stage of meiosis. The total population increases sixfold between days $14\frac{1}{2}$ and $17\frac{1}{2}$, reaches a slightly higher maximum value at day $18\frac{1}{2}$, and is reduced by about one-third, between day $18\frac{1}{2}$ *p.c.* and day 2 *p.p.* During this period four different waves of germ cell degeneration were distinguished.

In the rabbit ovary Chrétien (1966) found two main periods of germ cell multiplication, a first one between days 10 and 12 *p.c.* leading to a fourfold increase, and a second one between days 16 and 18 *p.c.* resulting in an approximately eightfold increase. The germ cell number reaches a maximum at day 26 *p.c.*, after which a marked decrease takes place by day 28 as a consequence of extensive degeneration. Since in the rabbit PGC migration starts at about day 10 and lasts at least until day 16 to 18, it is evident that PGCs actively multiply during migration. A similar germ cell history was found in the guinea pig by Ioannou (1964), in the monkey by Baker (1966), and in the human foetus by Baker (1963).

The most dramatic increase in germ cell number occurs in the ovary of the human foetus, viz. from 700–1300 during migration to approximately 600 000 in the second month of gestation, and further to a maximum of around 7 000 000 in the fifth month. After several waves of degeneration the number falls to about 2 000 000 at birth and further to approximately 300 000 at the age of seven. Since only about 500 eggs ovulate during the entire fertile period, there is a tremendous overproduction of female germ cells in human development (Baker, 1963).

Although in the human foetus oogonial multiplication reaches a maximum in the fifth month of gestation, Gaillard (1950) observed extensive germ cell formation in explants of ovarian cortex of 14-week to 36-week human foetuses and concluded that the potency for germ cell multiplication is not restricted to the early foetal period but certainly persists until birth. His explants first showed complete degeneration of parenchyma and young oocytes, after which new cord-like structures were formed from the 'germinal epithelium' which surrounds the explant. In these sex cords new oogonia appeared in large numbers. He left open the question of whether these oogonia originated from persisting but unrecognisable germ cells or were formed *de novo* from the somatic cells of the 'germinal epithelium'.

In the mouse Rudkin & Griech (1962) found that [³H]thymidine is incorporated into oocyte nuclei when administered half-way through the gestation period. The label was still found in the oocytes of female offspring at 6 weeks *p.p.*, suggesting that the PGCs present at the time of labelling had actually formed mature oocytes in the adult female. Borum (1966) found up to 100% labelling in the oocytes of female offspring after [³H]thymidine injections into pregnant mice at days 12 to 15 of gestation, but no incorporation into oocytes upon injection at different times after birth. He concluded that all oocytes present in the adult mouse have persisted from foetal life, so that at birth the mouse is furnished with a definitive population of oocytes from which all mature ova will later be derived. Peters & Crone (1967) showed that DNA synthesis occurs in oogonia in premeiotic interphase, the rate of synthesis falling during the foetal period in the mouse but within the neonatal period in the rabbit. Once synthesized the DNA persists in the growing and maturing oocytes. Kennelly & Foote (1966) concluded from [³H]methyl-thymidine injections into female rabbits on the day of birth and at 4 weeks *p.p.* that most, if not all definitive ova have already been formed at birth and that oocytogenesis *de novo* does not occur in the post-pubertal rabbit.

Unfortunately little is known about PGC numbers in forms which show a continuous or cyclic multiplication of the germ cells during the reproductive period, all the above forms belonging to the second variant of germ cell multiplication (see p. 74).

Spermatogenesis is usually characterised by either continuous or cyclic renewal of the germ cell population. In mammals it has been studied in recent years by Clermont & Leblond (1955, 1959), Roosen-Runge (1962), Leblond, Steinberger & Roosen-Runge (1963), Hilscher (1967), Roosen-Runge & Leik (1968) and Oakberg (1955, 1971), and reviewed by Clermont (1972), Hilscher & Hilscher (1976) and Roosen-Runge (1977). In these studies emphasis was placed on renewal of the spermatogonial stem cells and on the phenomenon of germ cell degeneration.

Hilscher & Hilscher (1976) state that when female and male gametogenesis in mammals are compared the 'gonia' stage of the female germ cells shows only a single proliferation wave, whereas that of the male germ cells shows a first proliferation wave, which is comparable with that of the oogonia, followed by a second wave after a preparatory interphase. The second proliferation wave is characterised by both stem cell renewal and differentiation into spermatocytes. According to Clermont (1972) the seminiferous epithelium of the mammalian testis is composed of five to six generations of germ cells. The spermatogonial population is renewed by a number of successive mitoses (three in man and up to six in the rat). Two types of spermatogonial stem cells can be distinguished, the type A 'reserve' and type A 'renewing' stem cells. The former type is not involved in the production of spermatocytes, while the latter type renews itself and simultaneously gives rise to differentiating spermatogonia at each mitotic cycle. However, the 'key' division between stem cell and differentiating spermatogonium does not have the character of a differential division in the morphological sense. The mechanism which regulates this choice is still unknown.

The general conclusion from the quantitative analysis of both oogenesis and spermatogenesis in the vertebrates is that there actually is no need to postulate any formation of germ cells *de novo* from somatic cells during any part of development or fertile life. It is therefore very likely that all the definitive male and female gametes descend from the initial population of germ cells which has arisen extra-gonadally during early embryonic development.

Experimental evidence

Elimination of the PGCs has been achieved surgically by removing that portion of the embryo in which the PGCs are located at the time of operation. It is a well-established fact that complete removal of the gonads leads to permanent sterility, but incomplete removal may lead to regeneration of the gonad and recovery of fertility.

In amphibians removal of the extra-gonadal source of the PGCs prior to their migration to the gonadal anlagen leads to the formation of gonads without PGCs and to permanent sterility. In neurulae of *Ambystoma mexicanum* Nieuwkoop (1947) removed the presumptive lateral plate meso-derm – the source of the PGCs in the urodeles – and obtained larvae which had more or less normal genital ridges but lacked PGCs. Vivien (1964) obtained similar results in *Lebistes* and *Xiphophorus* upon ^{32}P administration. Blackler (1962, 1965a) replaced the fertile region of the endoderm of *Xenopus laevis* neurulae – the source of the PGCs in the anurans – by a similar, more anterior region of the endoderm which does not contain PGCs, and obtained completely normal-looking tadpoles with normal gonadal anlagen but without PGCs. When reared to maturity the animals remained sterile.

In the anurans similar results were obtained by the elimination of the so-called 'germinal plasm', which is assumed to act as a germ cell determinant (see p. 87 for further discussion of this assumption). Pricking of the vegetal pole or UV-irradiation of the vegetal hemisphere performed at the 1- to 4-cell stage can lead to sterility of the gonadal anlagen of the larva and to subsequent permanent sterility. (For pricking of the vegetal pole, see Nieuwkoop & Suminski (1959), Fischiarolo (1960), Librera (1964), Buehr & Blacker (1970) and Gipouloux (1971). For UV-irradiation of the vegetal hemisphere, see Bounoure (1939), Bounoure, Aubry & Huck (1954), Padoa (1963a, b), Smith (1966), Tanabe & Kotani (1974) and Züst & Dixon (1975).)

It may therefore be concluded that destruction of the PGCs or their putative 'determinants' leads to permanent sterility. In other words, no formation of germ cells *de novo* occurs in animals which have no PGCs from an early stage of development. Although this seems a strong argument in favour of the continuity of the germ line, it is essentially negative and therefore inconclusive, for formation of germ cells *de novo* from somatic cells of the 'germinal' epithelium could depend upon a stimulating influence from existing viable germ cells (see Brambell, 1960). This rather unlikely postulate can only be disproved by heteroplastic or xenoplastic recombinations of presumptive germ cells and somatic gonadal tissue or by using specific nuclear markers. Such experiments were performed by Blackler & Fischberg (1961), who exchanged the fertile endoderm region between two strains of *Xenopus laevis* of which one contained the Oxford nuclear marker. Blackler (1962) made a similar exchange between two subspecies of *X. laevis*, while Blackler &

Gecking (1972*a*, *b*) performed the same experiment between *X. laevis* and *X. mülleri*, with subsequent intraspecific and interspecific matings. In all these experiments the germ cells showed the characteristics of the donor type that furnished the fertile endoderm region.

Similar experiments were carried out in urodeles by Smith (1964), who exchanged the ventro-lateral marginal zone (presumptive lateral plate mesoderm, representing the source of PGCs in the urodeles) between gastrulae of the white and the black axolotl. The successful cases in which the exchange was complete showed only progeny of the donor type, as tested by mating as well as histological examination. Because in Blackler's experiments the overlying ectoderm and mesoderm were also exchanged, while in Smith's experiments the graft not only furnished the source of the PGCs but also presumptive lateral plate mesoderm, it is possible, though rather unlikely, that the gonadal anlage, which is normally formed from the intermediate mesoderm, partly developed from graft tissue. Disregarding this minor objection, these experiments demonstrate that the presence of PGCs does not lead to a formation of germ cells *de novo* from the somatic components of the gonad. Moreover, no recent publications have led to any claim contradictory to the conclusions stated above (see also Blackler, 1965*a*, *b*).

The most elegant proof of the continuity of a single germ line was furnished in birds. First the extra-gonadal and extra-embryonic location of the PGCs in the so-called anterior germinal crescent of the blastoderm at early somite stages was demonstrated by surgical extirpation, cauterisation, UV-irradiation, γ-irridiation and X-irridiation. (Extirpation: Reagan (1916), Goldsmith (1935), Essenberg & Sreyda (1939), Simon (1960); cauterisation: Dantschakoff (1932*a*), Essenberg & Sreyda (1939); UV-irradiation: Benoit (1930), Reynaud (1968, 1969, 1970*b*); γ-irridiation: Dulbecco (1946), Marin (1959); X-irridiation: Dantschakoff (1933), Dubois (1962), Fargeix (1975).)

As will be discussed in chapter 7, in birds the PGCs are transported by the blood stream from the extra-embryonic germinal crescent to the gonadal anlagen. Reynaud (1969, 1970*a*, *b*, 1976*a*) was able to obtain repopulation of the gonadal anlagen after intravenous injection of a PGC suspension made from germinal crescent endoderm into a host embryo which had previously been sterilised by UV-irradiation of its germinal crescent. When he made xenoplastic recombinates of turkey PGCs with chick hosts or vice versa, he found that the gametes formed in the F1 generation were all of donor type. The colonisation of the host gonadal anlagen by foreign germ cells constitutes a crucial experiment, demonstrating that in birds the PGCs formed during early development are the sole precursors of the definitive gametes.

In mammals the evidence for the existence of a single and uninterrupted germ line is not as conclusive as in amphibians and birds, but is nevertheless fairly strong. X-irradiation (more than 168 r) of mouse gonadal anlagen after their colonisation with PGCs leads to the development of sterile gonads, which nevertheless show more or less normal proliferation of the gonadal epithelium (Everett, 1943). Mouse gonadal epithelium – prior to colonisation with PGCs – when grafted into the kidney capsule of host embryos remains sterile, whereas similar grafts made after colonisation with PGCs form typical testicular or ovarian tissues containing germ cells (Everett, 1943). These results plead strongly against any formation of germ cells *de novo*.

Further evidence comes from genetical studies. Mintz & Russell (1955, 1957) and Mintz (1957*a*, *b*, *c*) described several alleles of a mutation in mice called *dominant white spotting* (*W*), which in homozygous condition leads to sterility at birth. In the mutants the PGCs are formed at 8 days of gestation in the yolk sac endoderm in the normal number, which ranges from 10 to 100. The cells behave normally and migrate towards the gonadal anlagen. However, in normal mice the number of PGCs increases exponentially and reaches a value of approximately 5000 at the end of the migration period, whereas in the mutants the number of PGCs does not increase during migration. They subsequently degenerate, leading to total and permanent sterility of the gonadal anlagen. Twenty-five per cent of the offspring of heterozygous parents of the *W* series show the defect and are sterile at birth. This germ cell deficiency, which begins to manifest itself on the ninth day of embryonic development and is fully expressed at birth, furnishes a strong argument against any formation of germ cells *de novo* from the gonadal epithelium. This conclusion is further supported by culture *in vitro* of sterile mutant half-gonads fused with younger normal fertile half-gonads (Mintz, 1959, 1960*a*, *b*). In these chimaeric explants epithelial cells of the mutant gonad are in close contact with normal PGCs but no proliferation of mutant germ cells was ever observed. A similar germ cell deficiency seems to exist in the *steel* mutant (*Sl*) described by Bennett (1956). Radiation-induced damage (400 r) likewise results in a depletion of the PGC stock leading to permanent sterility. At lower doses the gonads may become repopulated, but all evidence points towards repopulation by remaining germ cells and not from somatic cells of the gonadal epithelium (Mintz, 1959, 1960*a*, *b*).

General conclusions

Summarising, it may be concluded that in the vertebrates the PGCs have an extra-gonadal origin during early embryonic development. The concept of a continuous, single germ line originating in early embryonic development has become highly probable for amphibians and birds and likely for mammals, while there is circumstantial evidence for it in agnathan fishes. Moreover, in recent years no compelling evidence against the general validity of this concept has been presented. Although further analysis is highly desirable, particularly in fishes and reptiles, it may be assumed for the present that in the vertebrates germ cell history is characterised by the one-time origin of

PGCs during early development and that the resulting PGC population gives rise to all the definitive gametes of the adult.

Waldeyer (1870) introduced the term 'germinal epithelium' for the putative germ-cell-producing outer layer of the gonadal anlage. The now convincingly demonstrated extra-gonadal origin of the PGCs makes this term, which is unfortunately currently used in the literature, a very confusing one (see also Roosen-Runge, 1977). It would be better to speak of 'gonadal epithelium' when describing the proliferative layer of the somatic gonadal anlage, and this is the term we shall use in the following chapters.

In the lower chordates the problem of the origin of the PGCs is still unsolved. As will be discussed in chapter 5 (p. 96), in the Cephalochordata the PGCs seem to arise *in situ* shortly before gonad formation. Moreover, gonadogenesis differs markedly from that in the vertebrates (see chapter 6, p. 108). Since the further history of the germ cells has not been studied, the conclusions drawn above for the vertebrates cannot yet be extended to the entire phylum of the chordates.

5

Site and mode of origin of the primordial germ cells

In discussing the site and mode of origin of the PGCs a distinction must be made between descriptive and experimental evidence. In the older literature the place where the PGCs were first identified during embryonic development was often taken as synonymous with their actual site of origin. We know now that the PGCs are formed extra-gonadally and have amoeboid motility, migrating through various tissues towards their final location in the gonadal anlagen. Although migration usually occurs during later stages of development (e.g. early larval stages), the possibility cannot be excluded that active displacements also take place during early stages. This may for instance be the reason for the often-observed accumulation of PGCs in the endo-mesodermal interspace in the Anamnia. Descriptive evidence may therefore only be considered as circumstantial.

It is clear that the earlier the stage of development at which the PGCs can be recognised, the more likely it is that the place where they are first identified corresponds to the actual site of origin. However, conclusive evidence for the site of origin of the PGCs requires experimental analysis. We may only learn about the causal factors responsible for germ cell formation (in other words, their mode of origin) from adequately planned experiments. These are virtually lacking in the lower chordates, and are mainly restricted in the Anamnia to the amphibians, and in the Amniota to the birds. The anuran and urodele amphibians will therefore be discussed first, followed by the various groups of fishes and the cephalochordates and tunicates. Among the amniotes the birds will be treated first, followed by the mammals and reptiles.

The Anamnia

Amphibia

Anura

The history of the germinal plasm. Since the discovery of the so-called 'germinal plasm' in the egg of *Rana temporaria* by Bounoure in 1931 and the elucidation of its subsequent history through embryogenesis, the anuran amphibians have served as the classical example of the presence of an uninterrupted germ line in the vertebrates (see Bounoure, 1939). These

81

observations agree very well with the results of similar studies in several insects and other invertebrates (for further information see: Everett, 1945; Cambar, 1956; Brambell, 1960; Blackler, 1970; Beams & Kessel, 1974; Eddy, 1975; Gipouloux, 1975; Smith & Williams, 1975). The correspondence in germ cell history, as well as the striking similarity in ultrastructure of the germinal plasm in groups of the animal kingdom which otherwise differ so much, has led to the idea that the presence of an uninterrupted germ line from the fertilised egg to the adult organism is a basic mechanism for maintaining organismal identity through successive generations. We must however test whether such a postulate is actually justified, and decide to what extent other mechanisms may also be represented in the chordates.

The fate of the germinal plasm during cleavage, gastrulation, neurulation and further embryogenesis was carefully studied for the first time by Bounoure in the years 1931 to 1934 (see Bounoure, 1939). In the vicinity of the vegetal pole of *Rana temporaria* eggs he found small subcortical patches of a special cytoplasm which he called 'germinal plasm' (cf. fig. 7.1, p. 114). During the first two cleavages these are more or less evenly distributed among the four blastomeres and are partially pulled inwards along the cleavage furrows. In subsequent cleavage cycles they are usually transferred to only one of the two daughter cells (differential division), and consequently are distributed among a restricted number of vegetal blastomeres. As a consequence of the pre-gastrulation movements the blastomeres containing germinal plasm are displaced together with other endodermal cells towards the centre of the vegetal yolk mass and sometimes even as far as the floor of the blastocoel. At late blastula (Bounoure, 1939) or early gastrula stages (Whitington & Dixon, 1975) the germinal plasm, which was hitherto situated at the periphery of the cell, is displaced intracellularly to a juxta-nuclear position, which is its final location in the PGC. Subsequently the PGCs are displaced along with the vegetal yolk mass by the gastrulation and neurulation movements, and are ultimately found among the vegetal blastomeres in the caudal portion of the embryo.

In the anurans a distinction must be made between true PGCs and their forerunners, the presumptive PGCs (pPGCs). During the first two cleavage divisions and the succeeding differential blastomere divisions one can only speak of pPGCs. One may define true PGCs as cells which are characterised by the presence of germinal plasm and which divide by clonal divisions. When the PGCs are first formed they do not divide very frequently, so that a considerable time may elapse between the last 'unequal' division and the first clonal division. It is interesting that this transition from presumptive to true PGCs coincides with the centripetal displacement of the germinal plasm. 'Unequal' divisions occur as long as the germinal plasm is situated peripherally, while 'equal' divisions take place when the germinal plasm has reached its juxta-nuclear position (Bounoure, 1939; Whitington & Dixon, 1975).

It is evident from the literature that in the anurans great significance is attributed to the germinal plasm for the determination and differentiation of the PGCs. However, what is the present evidence for this assumption? To test this question the following approaches have been made: (*a*) removal or destruction of the germinal plasm or its transplantation to an egg lacking functional germinal plasm, and (*b*) analysis of the chemical composition and possible function of the germinal plasm.

The significance of the germinal plasm for germ cell formation. As already described above, PGCs develop from blastomeres which contain germinal plasm. A number of workers have tried to remove the germinal plasm surgically. Nieuwkoop & Suminski (1959) were among the first to try to remove the germinal plasm at the 2- to 4-cell stage of *Xenopus* by superficial pricking of the vegetal pole region without causing a considerable outflow of cytoplasm. In general this did not lead to a marked reduction in the number of PGCs in the developing larvae. Fischiarolo (1960), using *Discoglossus* eggs found normal gonads with germ cells in metamorphosed larvae after exovation of about 40% of the egg contents as a consequence of pricking the vegetal region. Librera (1964) obtained varying results after withdrawal of about one-quarter of the egg contents with a micropipette inserted into the vegetal region of uncleaved eggs, 4-cell, blastula and early gastrula stages of *Discoglossus*. Of the 49 animals which reached metamorphosis, 32 had normal gonads, 5 had gonads with only a few germ cells, and 12 were completely sterile. Buehr & Blackler (1970) found either complete sterility or a marked reduction in the number of PGCs at the tadpole stage after pricking the vegetal pole at the 2- to 4-cell stage in *Xenopus*. They checked that some or nearly all of the germinal plasm was present in the exovate and found a quantitative relationship between the amount of germinal plasm remaining and the number of PGCs in the tadpole. From these experiments they concluded that the germinal plasm is indispensable for germ cell formation and that at least up to the tadpole stage germ cells are not formed from secondary sources. Gipouloux (1971, 1975) confirmed these results in *Discoglossus*, applying numerous superficial punctures to the vegetal pole region at the 2- to 4-cell stage. He ascribed the negative results of Nieuwkoop & Suminski to the experimental technique used, which did not lead to any substantial outflow of cytoplasm.

In 1937 Bounoure described for the first time the effect of UV-irradiation on the germinal plasm in the vegetal hemisphere of the fertilised anuran egg (see Bounoure, 1939). He observed a dramatic reduction (up to *c*. 90%) in the number of PGCs in the treated animals at metamorphosis. Although some animals had very few germ cells, he never actually obtained complete sterility. The gonads, though reduced in size, showed a normal somatic structure. Aubry (1953*a, b*) obtained still better results with UV-irradiation combined with animal–vegetal compression of the eggs. Padoa (1963*b*) studied UV sensitivity at different stages of development and found it to be highest in the period from shortly before the initiation of first cleavage till the beginning of second cleavage. He ascribed this to the accumulation of the germinal plasm in the vicinity of the vegetal pole shortly before first cleavage. The decrease in sensitivity after second cleavage seems to be due to the inward displacement of part of the germinal plasm along the cleavage furrows. This interpretation is supported by the observation that irradiation of the vegetal pole of the blastula no longer has any effect, since nearly all the germinal plasm is by now situated in the interior of the embryo. Smith (1966) obtained complete sterility in nearly 100 % of the developing larvae after UV-irradiation with 8000 erg/mm² of the vegetal hemisphere of *Rana pipiens* eggs at the beginning of first cleavage, whereas similar irradiation of the animal pole had no effect. By testing the effect of different wavelengths he obtained a UV spectrum for PGC inactivation. The maximally effective wavelength of 254 nm corresponds to the UV sensitivity of nucleic acids. Tanabe & Kotani (1974) also found complete sterility after irradiation of the vegetal hemisphere of the 2-cell stage of *Xenopus* with UV of wavelength 253.7 nm and dose *c.* 6000 erg/mm². When the stage of irradiation was gradually shifted from the 2- to the 4-cell stage an increasing number of PGCs was found in the larvae. When the dose of UV-irradiation at the 2-cell stage was decreased from 6000 to 750 erg/mm² the number of germ cells found in the tadpoles increased proportionally (see also Ijiri, 1976). The number of germ cells was also inversely proportional to the irradiated area.

In *Xenopus* Ikenishi, Kotani & Tanabe (1974) observed a swelling and vacuolation of the mitochondria and a fragmentation of the germinal granules as a direct effect of UV-irradiation of the germinal plasm. They concluded that maintenance of the normal structure of the germinal plasm is indispensable for its role as germ cell determinant, a conclusion supported by Ijiri (1977). Züst & Dixon (1975) and Clayton & Dixon (1975), however, warn against the conclusion that the effect of UV-irradiation represents a specific inactivation of the germinal plasm. They observed an inhibition of cleavage in the vegetal hemisphere in *Xenopus*, leading to a markedly abnormal segregation of the germinal plasm into particular blastomeres, and suggest that the sterilising effect of UV-irradiation is more likely to be due to a derangement of the normal cell lineage than to a direct deteriorating effect on the germinal plasm. Grant, Wacaster & Turner (1976) observed a disturbance of the cleavage process in the UV-irradiated vegetal half of the anuran egg and a subsequent delay in the gastrulation process.

The most elegant experiment indicating the great significance of the germinal plasm in PGC determination and differentiation was done by Smith (1966). He found that he could restore the germ line in *Rana pipiens* eggs previously sterilised by UV-irradiation by injecting cytoplasm from the vegetal pole of normal eggs, but that cytoplasm from the animal hemisphere was ineffective. However, whole vegetal cytoplasm was injected, *not* pure germinal plasm, so that this is not absolute proof of the indispensability of the germinal plasm. Nevertheless the restoration of PGC-forming capacity by injection of certain fractions of a crude homogenate of vegetal pole plasm into UV-irradiated *Rana* eggs suggests that germinal granules rather than mitochondria are indeed involved (Wakahara, 1977).

Very recently Dixon (personal communication) found that in contrast to all previous observations in *Xenopus* early PGC formation is probably not affected at all by UV-irradiation, but that the migration of the PGCs toward the genital ridges is markedly retarded, so that the gonadal anlagen become colonised only at a late stage of development. This finding, which requires further confirmation, seems to call for a reinterpretation of several of the results mentioned above.

A positive regulation after UV-irradiation towards the normal germ cell number at metamorphosis was noted by Bounoure *et al.* (1954). This may explain the negative results of Fischiarolo (1960) as well as the varying results obtained by Librera (1964). In Gipouloux's (1972) experiments negative regulation occurred, i.e. a restoration of the normal number of germ cells after an initial excess in a parabiont with one set of gonadal anlagen and two endodermal masses as sources of PGCs. Positive or negative regulation must be due to an increases or decrease in the number of mitoses in the PGCs between the tadpole stage and metamorphosis. Bounoure *et al.* (1954) actually observed a much higher mitotic rate in the PGCs of animals with a markedly reduced number of germ cells. The fact that Bounoure and Librera obtained metamorphosing tadpoles with only a few PGCs suggests that regulation may be impaired when the number of germ cells falls below a certain minimum level. This lack of regulation may, however, also be due to the remaining germ cells being damaged too much by the treatment used (see also Vannini & Gardenghi, 1964).

Composition and possible function of the germinal plasm. Gipouloux (1975) distinguishes three successive phases in germ cell formation. The first phase is characterised by the peripheral position of the germinal plasm during cleavage (pPGCs). In the second phase the germinal plasm occupies a juxta-nuclear position in contact with the nuclear membrane (PGCs). According to Gipouloux germ cell determination occurs in this phase, which is moreover characterised by an arrest of mitosis. The third phase is that of cellular differentiation of the PGCs in the gonadal anlagen.

Several authors have studied the composition of the germinal plasm in order to analyse its possible function in germ cell determination.

The UV sensitivity of the vegetal hemisphere of the anuran egg seems to correspond to that of nucleic acids (Smith, 1966). However, these nucleic acids are not necessarily in the germinal granules themselves, but may be those in the ribosomes and polysomes that are associated with the germinal granules at the unfertilised egg to early cleavage stages (Mahowald & Hennen, 1971), or even the mitochondrial DNA, the germinal plasm being very rich in mitochondria.

Smith (1965) investigated the capacities of germ cell and somatic nuclei for supporting normal development when injected into enucleated eggs of *Rana pipiens*. Eggs with germ cell nuclei showed a much higher incidence of cleavage than eggs with somatic nuclei (43% compared with 18%), while 40% compared with 0% of completely cleaved eggs formed normal embryos. He concludes that the nuclei of PGCs are apparently still totipotent, whereas the somatic nuclei have lost part of their potency. He therefore makes the suggestion that the germinal plasm in effect protects the nucleus from becoming channelled into a somatic pathway of differentiation (see also Nieuwkoop, 1964; Gipouloux, 1975).

Smith & Ecker (1970), reviewing the relevant literature, came to the conclusion that nucleic acids in the form of stable templates are present in the germinal plasm and act as germ cell determinants. Using autoradiography, Eddy & Ito (1971) failed to demonstrate any RNA synthesis in the electron-dense bodies of ovarian oocytes but found a rapid incorporation and turnover of amino acids. They also failed to detect any RNA cytochemically. They pointed out the great similarity between the electron-dense bodies found in oogonia and oocytes and the germinal plasm found in PGCs. Wallace, Morray & Langridge (1971), studying gene amplification during gameto-genesis, inferred the presence of self-replicating ribosomal DNA particles in the nucleolar core of oocyte nuclei. They suggest that during maturation these particles pass into the cytoplasm and end up in the germinal plasm. It must however be stressed that DNA has never been demonstrated in germinal plasm.

Deuchar (1972), who also considers a large store of ribosomal DNA as a characteristic of the germ line, assumes that during embryonic development and early larval life this DNA is transcribed into RNA, which is sequestered in the germinal plasm. Webb (1976) demonstrated ribosomal RNA synthesis associated with nuage material in oogonia. He suggests that the negative results of Eddy & Ito (1971) may be due to a protective effect resulting from the predominantly proteinaceous nature of the germinal plasm. Dziadek & Dixon (1977) found three phases of DNA synthesis in PGCs of *Xenopus laevis*, viz. in the stages 10 to 12, 22 to 24, and 37 to 39 (Nieuwkoop & Faber, 1975), no DNA synthesis occurring in the intervening periods. The PGCs distinguish themselves from surrounding endoderm cells as early as the mid-gastrula, since in the endoderm cells DNA synthesis is not temporarily arrested. In the PGCs RNA synthesis begins in the mid-gastrula (stage 11) and ceases at the early tail-bud stage (stage 24). Dziadek & Dixon did not find RNA synthesis in the germinal plasm and conclude that it is probably confined to the nucleus.

Züst & Dixon (1977) state that the germ cell lineage consists of a number of successive cell generations, so that the process of germ cell differentiation is a stepwise process. This may also hold for the process of germ cell determination.

Both the descriptive and the experimental evidence point to a crucial role for the germinal plasm in the determination and differentiation of the PGCs in the anuran amphibians. Only those blastomeres which contain a sufficient(?) amount of intact germinal plasm, or which receive it after previous UV treatment, develop into PGCs. Smith (1966) therefore calls the germinal plasm of the anuran egg *a germ cell determinant*. This means that the presence or absence of this plasm – more particularly of its nucleic acid component (Smith, 1966) – determines whether a cell will become a PGC or a somatic cell. The present authors feel, however, that one should be cautious in drawing such a far-reaching conclusion. So long as the germinal plasm or its active component has not been isolated and tested separately, Smith's conclusion must be considered premature. On the basis of the available evidence it can be said only that the vegetal pole region of the anuran egg contains factors which apparently act in germ cell determination. We shall return to this crucial question on p. 92 and at the end of this chapter (p. 102).

Urodela

Descriptive evidence. The descriptive evidence concerning the site of origin of the PGCs in the urodeles is somewhat scanty. On the basis of their yolk content Abramowicz (1913) concluded that in *Triturus* the PGCs arise from the endoderm and migrate into the mesoderm at a relatively advanced stage of development. In *Hemidactylium* and other urodeles, however, Humphrey (1925) identified the PGCs at an early tail-bud stage (from the 7-somite stage onwards) in the medial and later in the dorsal region of the lateral plate mesoderm. He stated that there is no evidence for Abramowicz's conclusion that their initial position is in the endoderm. Humphrey's observations were confirmed by Nieuwkoop in *Triturus* embryos at older tail-bud stages (unpublished) and recently in *Ambystoma* by Ikenishi & Nieuwkoop (1978), who could not, however, trace the PGCs back to stages earlier than stage 23 (H.).

Experimental evidence. Humphrey (1927, 1928, 1929) confirmed the localisation of the PGCs in the medial to dorsal lateral plate mesoderm by unilateral extirpation and by transplantation of this portion of the mesoderm to the ventro-lateral side of host embryos at stages 21 to 25. The former experiments led to the complete absence of PGCs on the operated side and the latter to additional germ cell formation. Nieuwkoop (1947) carried out a more systematic analysis of the origin of the PGCs in the urodeles. Neither removal of vegetal cytoplasm at the 1-cell stage, nor removal of the central portion of the vegetal blastomeres at the early gastrula stage in *Ambystoma* and *Triturus*, respectively, affected the number of PGCs in the developing larvae. Removal of the entire endoderm from early neurulae led to endoderm-free larvae containing a restricted number of PGCs situated in that portion of the mesoderm which had already invaginated and had apparently come into contact with the caudal endoderm prior to operation. Transplantation of the entire endoderm between neurulae of different *Triturus* species and between *Triturus* and *Ambystoma* showed that in the chimaeric embryos the PGCs always belonged to the species which had furnished the ecto-mesodermal envelope. Removal of the presumptive lateral plate mesoderm at the early neurula stage (compare fig. 5.1d) led either to complete sterility or to a very marked reduction in the number of PGCs, the remaining PGCs being situated in the unaffected, most cranial section of the fertile region. Heteroplastic transplantation of ventro-lateral marginal zone (presumptive lateral plate mesoderm) at an early yolk-plug stage ($11-11\frac{1}{2}$, H.) (compare fig. 5.1c) gave rise to chimaeric larvae in which the regional distribution of the PGCs of the two species corresponded to the regional composition of the chimaeric lateral plate mesoderm. These experiments demonstrated unequivocally that in the urodeles the PGCs are of mesodermal origin and are evenly distributed throughout the presumptive lateral plate mesoderm.

Contrary to Nieuwkoop, Takamoto (1953) observed complete absence of PGCs after removal of the entire endoderm from late neurulae of *Triturus pyrrhogaster*, but in Amanuma's (1958) experiments extirpation of the dorso-caudal endoderm of early neurulae of *T. pyrrhogaster* did not affect the number of PGCs. Amanuma (1957) moreover found that extirpation of presumptive lateral plate mesoderm at mid-gastrula or early neurula stages of *T. pyrrhogaster* and *Hynobius nebulosus* yielded larvae with a varying number of PGCs. He assumed that regulation occurred and concluded that the determination of the PGCs is not yet completed at the neurula stage. Orthotopic and heterotopic transplants of lateral plate mesoderm at tail-bud stages of *T. pyrrhogaster* by Asayama (1950, 1961) led to similar conclusions.

Lacroix & Capuron (1966) concluded from grafts of intermediate mesoderm of a tail-bud stage (stage 26, Gallien & Durocher, 1957) between Iberian and Moroccan races of *Pleurodeles* (which are characterised by a different lampbrush chromosome structure) that at this stage the PGCs are actually localised in the intermediate mesoderm. Maufroid & Capuron (1972) used defect experiments to determine the exact location of the PGCs in the lateral plate mesoderm at neurula and tail-bud stages of *Pleurodeles*, and found that in the early neurula the PGCs are located in the neighbourhood of the blastopore, from where they are displaced anteriorly (along with the lateral plate mesoderm) during mesodermal mantle formation. At tail-bud stages (stages 21 to 23, G. & D.) the PGCs move inside the lateral plate mesoderm

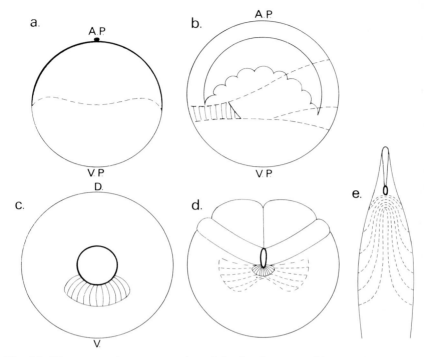

Fig. 5.1. Diagrammatic representation of the development of lateral plate mesoderm in the urodele embryo. (a) Distinction of animal and vegetal moieties in the uncleaved egg. (b) Location of presumptive lateral plate mesoderm in the ventral and latero-ventral marginal zone of the blastula. (c) External location of presumptive lateral plate mesoderm at yolk-plug stage. (d) Formation of left and right wings of· lateral plate mesoderm (solid lines uninvaginated, broken lines invaginated). (e) Cranial and ventral extension of lateral plate mesoderm at tail-bud stage; ventral view. A.P., animal pole; V.P., vegetal pole. (After Nieuwkoop, 1947.)

from a ventral and lateral position towards the vicinity of the Wolffian ducts. Ablation of lateral plate mesoderm containing PGCs did not lead to any formation *de novo* from adjacent mesoderm. The seemingly contradictory results of Asayama and Amanuma can be adequately explained on the basis of the exact location of the PGCs at successive stages of development. Maufroid & Capuron (1973) transplanted presumptive ventral and lateral marginal zone of early gastrulae of *Pleurodeles* into the blastocoel of host embryos of the same age, and observed the formation of extra ventral mesoderm with PGCs. They conclude that the PGCs are of mesodermal origin and are distributed throughout the entire ventral and ventro-lateral marginal zone at the early gastrula stage (see p. 88).

Asayama & Amanuma (1957) grafted an extra dorsal blastoporal lip into the ventral marginal zone of a host embryo, and concluded from the distribution of PGCs in the induced secondary anlage that PGCs are induced *de novo* in the secondary embryo. This conclusion seems to be at variance with the observation by Capuron (1968) that in *duplicitas* embryos the total number of PGCs never exceeded that of a normal embryo and was often considerably lower. This would indicate that the PGCs of the secondary embryo come from the same stock as those of the primary embryo and are not formed *de novo* from mesodermal tissues. However, the lateral plate mesoderm formation was not investigated in these *duplicitas* embryos.

Isolates of the ventral equatorial region of urodele blastulae, used as controls for mesoderm induction experiments (Sutasurya & Nieuwkoop, 1974), in a small number of cases formed some ventral mesodermal structures, which frequently contained PGCs. Maufroid & Capuron (1978) conclude from isolation and recombination experiments involving ventro-lateral marginal zone and ventral yolk endoderm that in *Pleurodeles* during gastrulation an additional inductive action by the endoderm plays a role in PGC formation. Similar experiments with ventral marginal zone by Nieuwkoop & Sutasurya (unpublished) suggest, however, that in *Ambystoma* PGC formation does not require an additional inductive action by the endoderm. Here PGC determination seems to occur very early in development, the PGCs constituting one of the self-differentiation pathways of the ventral mesoderm see below, p. 91). These contradictory results show that in the urodeles the time and mode of determination of the PGCs are still an open question.

UV-irradiation of the vegetal hemisphere of the *Pleurodeles* egg has no effect on PGC formation when performed at cleavage, blastula or early gastrula stages, but leads to a dramatic reduction of the number of PGCs or even to complete sterility when applied at a late yolk-plug stage (Capuron, personal communication). At that stage the presumptive ventro-caudal mesoderm – the source of the PGCs – is localised in the ventral blastoporal lip and is therefore accessible to the irradiation. In our opinion this result indicates that the PGCs pass through a sensitive period at the late gastrula stage, which possibly represents a (first) decisive step in their determination.

Attention must now be drawn to a different line of approach. Kotani (1957) replaced presumptive lateral plate mesoderm of early gastrulae of *T. pyrrhogaster* by presumptive epidermis and found PGC formation in the grafts. He concluded that the presumptive epidermis of early gastrulae possesses germ-cell-forming potencies. Smith (1966) replaced ventro-lateral marginal zone of stage 11 gastrulae by presumptive epidermis of early gastrulae, using white and black axolotls. When testing the progeny of the operated embryos, he did not find PGC formation of graft origin. Kotani's results were later confirmed by Kocher-Becker & Tiedemann (1971), who found PGC formation in partially mesodermised and endodermised 'ectodermal' explants from *Triturus* early gastrulae treated with a vegetalising factor isolated from chick embryos.

In the course of the analysis of mesoderm induction in the urodeles (see chapter 2, p. 10), and Nieuwkoop, 1969*a*, *b*, 1970, 1973; Nieuwkoop & Ubbels, 1972), Boterenbrood & Nieuwkoop (1973) found that PGCs are a normal constituent of induced ventral mesoderm. When studying the mesoderm-inducing capacities of the dorsal, lateral and ventral portions of the yolk mass they found that both the lateral and ventral portions induce predominantly ventral mesoderm, while the dorsal portion induces massive dorsal, axial mesodermal structures. When comparing the inductive capacities at various stages in *Ambystoma* they found that by a very early gastrula stage (stage 10$^-$, H.) the mesoderm-inducing capacity of the dorsal endoderm has virtually vanished, that of the lateral endoderm is already markedly reduced, and that of the ventral endoderm is slightly diminished. This renders it very likely that the entire endoderm will have lost its mesoderm-inducing capacity by stage 11; this may explain the contradictory results of Kotani and Smith, who for their grafting experiments used early and middle gastrula stages, respectively. In *Pleurodeles*, however, the germ-cell-inducing capacity of the ventral endoderm seems to persist during gastrulation (Maufroid & Capuron, 1977). Kotani's conclusion is corroborated by the study of Sutasurya & Nieuwkoop (1974), who found that under the inductive influence of ventral endoderm PGCs can be formed from any part of the totipotent animal moiety of the urodele blastula, the competence being highest in the 'ectoderm' near the equator and lowest at the animal pole. In normal development PGC formation is restricted to the ventral and ventro-lateral marginal zone by an inhibiting influence from the presumptive notochord spreading dorso-ventrally through the mesodermal mantle. From these experiments Sutasurya & Nieuwkoop conclude that in the urodeles PGCs *do not develop from predetermined elements, but arise strictly epigenetically from common, totipotent cells of the animal moiety of the blastula* as part of the regional induction of the mesoderm by the vegetal yolk endoderm.

When comparing PGC formation in the urodeles with that in the anurans, one is unavoidably led to the conclusion that not only do the PGCs originate from two different sites in the two groups, but that there are moreover two fundamentally different mechanisms at work. (This conclusion is clearly in flat contradiction to a statement by Blackler (1968) to the effect that the different site of origin of the PGCs in anurans and urodeles is only a minor variation and has no development significance.) In the anurans all the PGCs originate from the endodermal moiety of the egg in the vicinity of the vegetal pole, whereas in the urodeles they arise from the animal 'ectodermal' moiety, more particularly the presumptive lateral plate mesoderm in the ventral to ventro-lateral equatorial region. In the anurans all the descriptive and experimental evidence pleads in favour of the *predetermined* nature of the pPGCs, based on the presence of a germ-cell-specific cytoplasmic component, the 'germinal plasm', which is present in the embryo from the very beginning of development. In contrast, in the urodeles the PGCs develop strictly epigenetically from common, totipotent cells of the animal moiety under the inductive influence of the ventral yolk endoderm.

Anuran and urodele PGCs seem to have one characteristic in common, i.e. their irreplaceability. Removal of the region from which they originate leads to sterility of the developing larvae in both groups. Regulatory formation of new PGCs from adjacent regions apparently does not occur. This could lead to the conclusion that the PGCs are already determined or at least partially determined at the stage of operation. As already mentioned on p. 90, in the urodeles the PGCs indeed seem to be determined very early in development and probably constitute one of the self-differentiation pathways of the mesoderm in the ventral marginal zone region.

The most striking difference in the mode of origin of the PGCs between anuran and urodele amphibians seems to be the role of the germinal plasm in germ cell determination and differentiation. In chapter 3 it was shown that the presence of germinal plasm is also a characteristic of urodele PGCs. Although ultrastructurally it looks identical to that of the anuran PGCs, it appears only at a late stage of development, i.e. when the cellular differentiation of the PGCs has already begun and thus long after their actual determination must have taken place. In the urodeles the germinal plasm cannot therefore play the role of germ cell determinant (Ikenishi & Nieuwkoop, 1978). It is very difficult to conceive that in one animal group a certain organelle acts as a determinant of a special cell type whereas in another group it constitutes only an attendant structure, being a differentiation product of that same cell type! This incongruity raises the question of whether in the anurans the characteristic structures of the germinal plasm actually do act as germ cell determinant, or whether another, non-structural component of the cell is responsible for germ cell determination. (See Aisenstadt, 1975.)

Coecilia

There is only one reference, Marcus (1938), on the possible site of origin of the PGCs in the Coecilia, the most primitive group of recent amphibians. In histological sections of rather late embryonic stages of *Hypogeophis* Marcus identified the PGCs in the endoderm in the vicinity of the blastopore. At later stages he found them in the dorsal mesentery and concluded that they had arrived there either direct from the dorsal endoderm or by passing through the lateral plate mesoderm. They are finally located in the genital ridges ventral to the Wolffian ducts. Marcus states that the PGCs have an 'endodermal' appearance. The evidence is inconclusive, however, the more so since it is known that PGCs may migrate from one tissue into another at relatively early stages of development and may engulf yolk platelets from cytolysing cells.

The fishes

As regards the fishes a clear distinction must be made between the development of the four classes: the Agnatha or cyclostomes, the Chondrichthyes or elasmobranchs, the Osteichthyes, and the Teleostomi or teleosts.

As we have seen in chapter 2 (p. 22), the early development of the eggs of the Agnatha and Osteichthyes, which both represent ancient groups of fishes, is rather alike and shows close parallels with the early development of the amphibians. The development of the yolk-rich egg of the Chondrichthyes is rather different from that of the former groups, while the Teleostomi constitute a highly specialised group with a very aberrant development.

Agnatha

The only data available at present on the origin of the PGCs in the Agnatha are those of Okkelberg (1921) on the brook lamprey. He found PGCs first at an early tail-bud stage, in the most caudal portion of the trunk where they are scattered in the lateral plate mesoderm (fig. 5.2). Later they accumulate in the vicinity of the Wolffian ducts. This descriptive evidence suggests a mesodermal origin of the PGCs but is inconclusive.

Osteichthyes

On the origin of the PGCs in the Osteichthyes only descriptive data from a few species are available. In *Amia* some PGCs are found in the peripheral region of the endoderm at an early larval stage, but the majority are present in the adjacent lateral plate mesoderm (Allen, 1911). In *Lepisosteus* Allen (1911) found them in the ventral and lateral portions of the single-layered gut endoderm. Reinvestigating the original material of *Polypterus* collected by Kerr, De Smet (1970) found PGCs were first visible in an early larval stage among the yolk-laden cells of the roof of the gut, at the boundary between the gut endoderm and the lateral plate mesoderm and in the lateral plate mesoderm itself. Finally, in *Acipenser* Maschkowzeff (1934) was first able to detect PGCs in the intestinal endoderm. On the basis of the yolk content of the PGCs the various authors concluded that in all the four species studied they are of endodermal origin. They are supposed to migrate from their endodermal site of origin towards the lateral plate mesoderm and then to the dorsal midline, where they accumulate near the nephrotomes. It must however be said that the available descriptive evidence suggesting an

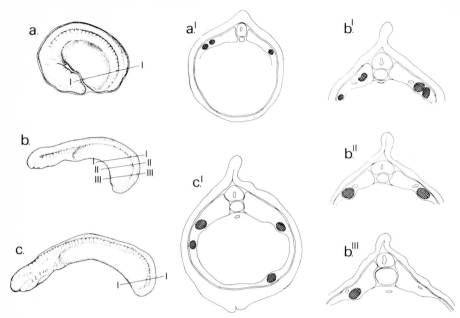

Fig. 5.2. Location of PGCs (hatched areas) in the brook lamprey embryo at three successive tail-bud stages (a–c), with cross sections through the caudal trunk (a[1], b[1]–b[111] and c[1]). (Redrawn from Okkelberg, 1921.)

endodermal origin of the PGCs in the Osteichthyes is very scanty and inconclusive.

Chondrichthyes

Data on the origin of the PGCs in the elasmobranchs are similarly scanty and inconclusive. In *Raja batis* Beard (1900, 1902*a, b*) found them for the first time at an open neural plate stage. The majority were located in the peripheral dorso-lateral endoderm, some in the adjacent lateral plate mesoderm, and even a single one in the overlying ectoderm. In older embryos they were found in the splanchnic mesoderm or between this and the intestinal epithelium. Beard (1920*a*) found a similar location of the PGCs in *Raja radiata, Pristiurus melanostomus, Scyllium vulgaris, Torpedo ocellata* and *Acanthias vulgaris*. This descriptive evidence points towards a relatively early and endodermal origin for the PGCs in the Chondrichthyes, but is again far from conclusive.

Teleostomi

Eigenmann (1891) concluded from the size of the PGCs in *Micrometrus* (= *Cymetogaster*) that they must have segregated from the somatic cells at an early cleavage stage. In *Fundulus* Richards & Thompson (1921) were first able to identify the PGCs in the peripheral endoderm in 46- to 53-hour embryos (early tail-bud stages). They stressed the close association of the PGCs with the syncytial periblast, an observation which was confirmed by Hann (1927) and Wolf (1931). In *Scardinius* Reinhard (1924) described their possible segregation from the syncytial periblast. In the salmonid *Micropterus* Johnston (1951) identified the PGCs first in an excrescence of the periblast protruding over the caudal edge of the blastodisc (erroneously called the dorsal blastoporal lip). At a slightly later stage they were found in the caudal (erroneously called ventral) periblast, from where they moved into the caudal extension of the yolk sac. From there they wedged their way between the splanchnic mesoderm and the gut endoderm towards the mid-dorsal region, and subsequently towards the genital ridges. In *Oryzias* PGCs were first found in the unsegregated mesendoderm, occasionally also in the ectoderm, at an 'early gastrula stage' (Gamo, 1961*b*). After the segregation of mesoderm and endoderm they were found in all three germ layers. Since their number increased in the absence of detectable mitotic divisions, Gamo assumed that they segregated during the entire early development. In early stages they were closely associated with the syncytial periblast. After myotome formation they were found predominantly in the peripheral endoderm, where they seemed closely associated with yolk protrusions. Gamo also noticed the capacity of the PGCs to absorb yolk. Only those PGCs which are incorporated into the gonadal anlagen become definitive germ cells, while the ectopic germ cells degenerate. According to Pala (1970), in *Gambusia* PGCs are first discernible among the deeper cells of the embryonic node in the caudal half of the blastodisc at an 'early gastrula stage'. During the further segregation of the embryonic anlage they are found in the mesodermal portion of the mesendoderm. Nedelea & Steopoe (1970) localised the PGCs in the mesodermal layer of the embryonic shield of *Cyprinus* shortly before the completion of the overgrowth of the yolk.

The only experimental evidence available (Oppenheimer, 1959*b*) seems to argue for a mesodermal origin of the PGCs. Randomly reassociated fragments of posterior thirds of embryonic shields of 'middle gastrulae' (about half way in the overgrowth of the yolk), grafted into the extra-embryonic area of a host embryo, formed gonads with PGCs in the absence of any endodermal structures.

Summarising, it must be said that there is no concordant evidence for a particular site of origin of the PGCs in the teleosts (see also the review by Vivien, 1964). The often-observed close association of the PGCs with the syncytial periblast may well be of a secondary nature in view of the tendency of the PGCs to associate themselves with a nutritional source. Nothing can at present be said about the mode of origin of the PGCs.

The lower chordates

Tunicata

Except for the Larvacea, which are sexually mature tadpole-like organisms, all the other tunicates undergo metamorphosis, during which the axial structures (notochord and somites) so characteristic of the chordates are resorbed. As far as is known PGCs are only found in metamorphosed, sessile individuals.

Berrill (1975) states that among the tunicates various forms of asexual reproduction are found, during which a small fragment of the parental body is set apart for the formation of a new zooid. In several forms sexual and asexual generations alternate, and in these forms no germ line is therefore traceable. The germ cells in the hermaphroditic gonads – the ovotestes – seem to arise from small masses of undifferentiated mesodermal cells. These cells are called haemoblasts, suggesting an origin from the vascular system.

In summary it may be said that in the tunicates the PGCs probably originate in the mesoderm, although their mode of origin is unknown.

Cephalochordata

According to Boveri (1892) the PGCs in *Branchiostoma* larvae are found in clusters at the ventro-cranial extremities of the eleventh to thirty-sixth of the 61 somites (fig. 5.3). The '*Grenzzelle*' which Hatschek (1888) found at the ventral extremity of each somite in a younger larva may represent the segmental stem cell of the PGCs. The strictly segmental origin of the PGCs pleads in favour of their differentiation *in situ* from the intermediate mesoderm.

The Amniota

Aves

Swift (1914) was the first to locate the PGCs at early somite stages, in an anterior, extra-embryonic crescent-shaped area at the boundary between the area pellucida and the area opaca, which he called the 'germinal crescent' (see fig. 7.3b, c, p. 119). In the next decade this observation was both confirmed and contradicted, but it gradually found wider approval. Willier (1937) found PGCs in chorioallantoic grafts of portions of the chick blastoderm containing part of the germinal crescent, whereas grafts without this

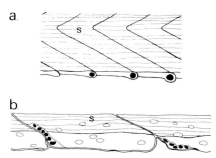

Fig. 5.3. Lateral view of part of a row of somites (s) in larva of *Branchiostoma lanceolatum.* (a) Segmental location of presumed stem cells of PGCs (black) in early larva (Hatschek's '*Grenzzelle*', 1888). (b) Similar arrangement of clusters of PGCs in slightly older larva. (After Boveri, 1892.)

region were sterile. Simon (1960) observed complete sterility or a very marked reduction in the number of PGCs after removal of the germinal crescent. Reynaud (1969) destroyed all the PGCs by UV-irradiation of the germinal crescent, and was able subsequently to restore a normal PGC population in the irradiated embryo by intravenous injection of a cell suspension made from the germinal crescent of another embryo.

The spatial extension of the germinal crescent apparently varies markedly in different species and breeds. In the chick and turkey it usually surrounds only the anterior half of the embryonic anlage (Simon, 1960; Reynaud, 1969; Clawson & Domm, 1969; Ukeshima & Fujimoto, 1975), but in the quail it extends around the anterior two-thirds (Reynaud, 1976*b*), while in a chick hydrid and in the duck its extension is even more variable and may reach the posterior side of the embryonic anlage (Fargeix, 1969; Bruel, 1973).

The PGCs are initially located in the endodermal hypoblast layer of the germinal crescent (Clawson & Domm, 1969). They then segregate from the hypoblast and accumulate at the junction of the ectoderm and endoderm. They begin to enter the extra-embryonic blood islands as soon as the invaginated mesoderm reaches the germinal crescent area. This has recently been confirmed by Fujimoto *et al.* (1976*a*) and others.

Although the extra-embryonic location of the PGCs at early somite stages is by now well established, little is known about the *primary origin* of the PGCs and the actual time of their determination.

Quite a few experiments have been done at the unincubated blastoderm stage. Dubois (1967) concluded from in-vitro culture of parts of the unincubated blastoderm that the PGCs originate from the primary hypoblast in the posterior region of the blastoderm and are transported anteriorly by its morphogenetic movements (see also Vakaet, 1970). Dubois later found the PGCs at the time of oviposition to be distributed in a transverse band in the posterior half of the embryonic anlage (see Dubois & Croisille, 1970). This seems to be at variance with the findings of Fargeix (1969), who obtained a 'normal' number of PGCs (i.e. equal to that of a normal embryo) from anterior halves of unincubated blastoderms, while posterior halves formed either a 'normal' or a deficient number of PGCs. Lateral halves always formed a 'normal' number of PGCs. The number of PGCs decreased in cultured posterior halves as they were isolated at later stages. Rogulska (1968) found that duck blastoderms transected *in vivo* formed double the 'normal' number of PGCs. In those cases where the embryonic anlagen were oriented parallel to each other, the PGCs were predominantly located in the zone between them. Working with freshly laid chicken 'winter' eggs which did not yet show hypoblast formation, Eyal-Giladi, Kochav & Menashi (1976) found that both anterior and posterior halves formed the 'normal' number of PGCs.

Fargeix (1975) showed that X-irradiation of the entire blastoderm can lead ultimately to the sterility of the embryo. In such embryos the number of PGCs is unaffected at an early somite stage but decreases later, whereas in unirradiated embryos the number of PGCs increases steadily. Fargeix found that in the unincubated blastoderm irradiation of either the anterior or the posterior half led to the same reduction in PGC number. However, irradiation of the posterior half of the blastoderm incubated for 10 hours led to only a slight reduction in the number of PGCs, whereas a marked reduction was obtained after the same treatment of the anterior half. From these experiments he concluded that the pPGCs are evenly distributed throughout the unincubated blastoderm, but accumulate in the anterior half during the first hours of incubation. This is in accordance with the segregation *in situ* of the primary hypoblast from the epiblast (Eyal-Giladi & Kochav, 1976) and its subsequent anterior displacement as a result of the expansion of the secondary hypoblast from Koller's sickle (Vakaet, 1962).

The formation of a 'normal' number of PGCs from anterior and posterior halves of unincubated blastoderms seems to be at variance with the assumption that predetermined elements are present in the primary hypoblast – unless a mechanism exists which leads to a doubling of the number of PGCs after transection. In this connection two contradictory observations are of interest. Fargeix (1969) states that transected blastoderms are always retarded in development, so that there would seem to be enough time for an extra clonal division of the PGCs. In contrast Swartz & Domm (1972), who studied mitoses in PGCs by colchicine treatment, observed that the clonal divisions of the PGCs start late, i.e. not before the migration period.

A possible alternative explanation for the 'normal' number of PGCs in half embryos, first suggested by Fargeix (1969) and again by Eyal-Giladi *et al.* (1976), is that predetermined elements in the sense of the anuran pPGCs do not exist in the early development of birds, but that (in both half and normal embryos) the PGCs are induced (in 'normal' number) at a later stage of development.

If an early segregation of pPGCs does occur in birds, it nevertheless seems to differ from that in the anuran amphibians, since no germinal plasm has hitherto been demonstrated in birds. The prime characteristic of avian PGCs, i.e. their high glycogen content, cannot be regarded as a germ cell determinant since it is not specific to PGCs and appears relatively late in development. It probably only represents a condition for their migratory activity. In birds the role of germ cell determinant cannot at present be attributed to any particular ultrastructural or physiological characteristic of the PGCs.

Reynaud (1968, 1969, 1970a, b) observed accelerated multiplication of the germ cell population re-established in previously sterilised chick hosts by injection of a small number of turkey PGCs. Lutz & Lutz-Ostertag (1972) called this phenomenon auto-regulation. Reynaud (1971a) studied 'normal' germ cell multiplication in chick, quail and turkey. Auto-regulation was also observed by Fargeix (1976) in X-irradiated anterior and posterior halves of unincubated duck blastoderms.

Summarising, it may be said that the early segregation of the avian PGCs from the primary hypoblast shows certain similarities to the history of the germ cells in the anuran amphibians. However, the avian PGCs are not characterised by the presence of germinal plasm or any other germ-cell-specific structure. The suggestion by Fargeix (1969) and by Eyal-Giladi *et al.* (1976) that avian PGCs may be formed epigenetically during embryogenesis (as they are in the urodeles) cannot be excluded, but does not seem to be easily reconciled with the endodermal origin of the PGCs and with the results of X-irradiation at early blastoderm stages. It must therefore be concluded that the actual mode of origin of the PGCs in birds is still unknown.

Mammalia

The data relevant to the site and mode of origin of the PGCs in the mammals come predominantly from the analysis of the early development of the mouse embryo, to which we will therefore mainly refer.

Brambell (1927) was the first to recognise the extra-gonadal origin of the PGCs when he detected them in the extra-embryonic yolk sac endoderm of the 8-day mouse embryo. During gut formation they are displaced towards the hindgut endoderm. Leaving the endoderm, they migrate through the dorsal mesentery towards the genital ridges. This observation was confirmed by Everett (1943). Chiquoine (1954), on the other hand, stated that in the 8-day mouse embryo the PGCs originate from the yolk sac splanchnic mesoderm and perhaps even from the caudal portion of the primitive streak, thus suggesting a mesodermal origin. In 1960 Brambell again inferred an endo-dermal origin for the PGCs. Mintz (1960a) observed them first in the yolk sac endoderm in the 8-day presomite mouse embryo (see fig. 7.4, p. 124). Blandau, White & Rumery (1963) found the PGCs at the base of the allantoic evagination. The endodermal origin of the PGCs was again questioned by Ozdzenski (1967), who found them in the caudal region of the primitive streak as well as in the anlage of the allantois of presomite embryos. He concluded that they migrate from the primitive streak mesoderm into the endodermal layer. Bruel-Beaudenon & Hubert (1968) speak of a 'posterior germinal crescent' in the mammals. Hamilton & Mossman (1972) again have the PGCs originating from the extra-embryonic yolk sac endoderm. Merchant & Zamboni (1973) identified the PGCs in the gut epithelium of the 9-day mouse embryo. They came to the conclusion that the PGCs associate themselves with other tissues because as the result of insufficient yolk material they need an exogenous energy source, the alkaline phosphatase activity at the cell periphery denoting an active intercellular exchange of material. The location of the PGCs in the yolk sac endoderm of the 8- to 9-day mouse embryo was confirmed by Spiegelman & Bennett (1973). Clark & Eddy (1975) agree with Merchant & Zamboni (1973) that the PGCs reside only temporarily in various somatic tissues during their life history, and depend upon the somatic tissues for their nutrition. Although they found the PGCs for the first time in 8- to 8½-day embryos in that portion of the yolk sac endoderm that later forms the hindgut, the PGCs do not resemble the surrounding endodermal cells in their ultrastructure but look much more similar to the adjacent mesodermal cells. They therefore propose that the PGCs do not in fact originate from the yolk sac endoderm but from the mesoderm.

In other mammalian species the PGCs are likewise first detected in the extra-embryonic yolk sac endoderm and the allantoic endoderm and mesoderm of presomite embryos (Celestino da Costa (1932a, b) and Gomes-Ferreira (1956, 1957) in the guinea pig; Witschi (1948), Asayama (1965) and Fujimoto et al. (1977) in the human). At slightly older stages they are found in the hindgut endoderm (Celestino da Costa (1937) in the guinea pig; Chrétien (1966) in the rabbit; Sawano (1959) in the human).

The only disparate observation concerns a study by Vanneman (1917) in the marsupial *Armadillo*. In this polyembryonic form (see chapter 2, p. 47) the PGCs were identified for the first time between the extra-embryonic ectodermal and endodermal layers, outside the primary embryonic buds. This more or less resembles the situation in birds, but the finding must be viewed with caution since it has never been confirmed.

For the origin and continuity of the germ line in mammals the reader is further referred to the reviews by Mintz (1960a, b), Pasteels (1964) and Peters (1970).

The only experimental evidence comes from Gardner & Rossant (1976), who produced chimaeras for coat colour genes in the mouse by injecting embryonic ectoderm cells into host blastocysts, and found that PGCs were formed from the injected cells. They conclude that the PGCs originate from (totipotent) precursor cells of the definitive embryo rather than from the extra-embryonic yolk sac endoderm. The totipotent embryonic 'ectoderm' would yield both PGCs and somatic cells, the segregation of which would occur only later. This situation rather resembles that in the urodele amphibians where the lateral plate mesoderm, from which the PGCs originate, arises in the process of mesoderm formation from the totipotent animal 'ectodermal' moiety of the blastula. However, due to our very restricted knowledge of the process of mesoderm formation in the mammalian embryo (cf. chapter 2, pp. 47–49) we can only signalise the parallel.

Recently Falconer & Avery (1978) concluded that in the mouse the variability observed in chimaeras produced by aggregation of cleavage stages and in mosaics resulting from X chromosome inactivation provides additional evidence that the PGCs originate in the primary ectoderm and not in the yolk sac endoderm.

Reptilia

The origin of the PGCs in the reptiles has been traced back no further than the early somite stages. In all reptiles studied the PGCs were found in the extra-embryonic area of the blastoderm. As to the more precise location of the PGCs the reptiles fall into two categories: (*a*) those with an anterior germinal crescent, which may extend around the entire embryonic anlage, and (*b*) those with a posterior germinal crescent. The former localisation more or less resembles the situation in birds, the latter that in mammals (Bruel-Beaudenon & Hubert, 1968). A strictly anterior germinal crescent was found in the snake *Vipera* by Hubert (1969), while an anterior localisation with extensions to the lateral and posterior sides seems to exist in the Lacertilia *Mabuya* and *Chamaeleo* (Pasteels, 1953) and the slow-worm *Anguis* (Hubert, 1971*b*). Both anterior and posterior germinal crescents were described in *Sphenodon* by Tribe & Brambell (1932). A strictly posterior germinal crescent was found in the chelonians (Allen, 1906; Risley, 1933) as well as in several *Lacerta* species (Dufaure & Hubert, 1965) (fig. 5.4). It is evident that the two types of localisation do not coincide with particular taxonomic groups, since the Lacertilia show both types. According to Hubert (1976), in the forms having an anterior germinal crescent (e.g. *Mabuya*, *Chamaeleo*) the PGCs are displaced anteriorly together with the primary hypoblast by the gastrulation movements, whereas in those having a posterior germinal crescent (e.g. the chelonians and *Lacerta*) the localisation of the PGCs is not affected by the gastrulation process, which occurs more anteriorly.

At the stages examined the PGCs were found in the endodermal layer before the mesoderm has spread between the ectoderm and endoderm (Allen, 1906; Tribe & Brambell, 1932; Risley, 1933; Pasteels, 1953). According to Hubert

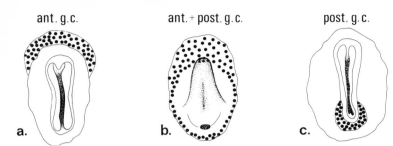

ant. g. c.   ant. + post. g. c.   post. g. c.

a.   b.   c.

Fig. 5.4. Different extra-embryonic locations of PGCs in reptiles. (a) Anterior germinal crescent (ant.g.c.) in several Lacertilia and Ophidia. (b) Both anterior and posterior germinal crescent (ant. + post.g.c.) in *Sphenodon*. (c) Posterior germinal crescent (post.g.c.) in Chelonia and some Lacertilia.

(1976) the PGCs originate either in the primary hypoblast or in the definitive hypoblast formed during gastrulation. For general reviews the reader is referred to Pasteels (1964) and Dubois (1965).

The only experimental evidence available, which was obtained by the extirpation of the posterior germinal crescent in *Lacerta vivipara* either before or after the migration of the PGCs towards the genital ridges (Hubert, 1971*a*), confirms the descriptive data.

Summarising, it must be concluded that little is known about the actual site of origin of the extra-embryonic PGCs in the reptiles. The observed localisation in the anterior or posterior germinal crescents probably represents a secondary situation. About the mode of origin of the PGCs in the reptiles we are entirely in the dark.

General conclusions

Surveying the entire phylum of the Chordata with respect to the site and mode of origin of the PGCs, it is evident that at least two essentially different mechanisms are involved in individual groups, viz. (*a*) a very early segregation of the germ line from the somatic cell lines, and its determination by a special cytoplasmic component, and (*b*) a strictly epigenetic development of the germ cells from totipotent embryonic cells under an inductive influence. These two alternative mechanisms correspond to an endodermal and a mesodermal origin of the PGCs, respectively. The existence of the two mechanisms has been deduced from the experimental analysis performed in anuran and urodele amphibians. The late appearance of the germinal plasm in the urodele PGCs moreover throws doubt on the determinative role of the germinal plasm generally.

When surveying the chordates we must unfortunately state that it is at present only in the two groups of amphibians where the situation is sufficiently clear. Our knowledge of the development of the urochordates, in which as far as we know the PGCs are only detectable after metamorphosis, does not allow any conclusion. In the cephalochordates the PGCs also seem to arise rather late in development, i.e. during the final differentiation of the mesoderm. Their clearly segmental localisation ventral to the somite segments pleads in favour of differentiation *in situ* within the epigenetically formed mesoderm, suggesting a 'urodele' type of germ cell formation. In the Agnatha the PGCs are first found scattered in the lateral plate mesoderm at an early tail-bud stage, a situation which again resembles the 'urodele' type of germ cell formation. In the Osteichthyes there is some purely descriptive evidence for an endodermal localisation of the PGCs, but this does not say anything about their actual site and mode of origin. The same uncertainty holds for the situation in the Chondrichthyes. The purely descriptive evidence regarding the Teleostomi is far from concordant and does not allow a decision one way or the other. No experimental evidence is available for the Coecilia.

When considering the amniotes, the Reptilia are almost unknown terrain, apart from some evidence that the situation in the Chelonia differs from that in the Squamata. In contrast much experimental evidence is available on the birds. The PGCs seem to originate in the extra-embryonic primary hypoblast during early development. The lack of germinal plasm in avian PGCs possibly means their actual mode of origin is different from that in the anuran amphibians. It also calls into question once more the determinative role of the germinal plasm. In the mammals the still scanty experimental evidence suggests an epigenetic mode of origin of the PGCs, in line with the 'urodele' type of germ cell formation.

6

The formation of the gonadal anlagen

Our knowledge of the formation of the gonadal anlagen in the chordates comes chiefly from the amphibians and birds. We shall therefore start with the amphibians, then discuss the situation in the fishes and lower chordates, subsequently consider the birds, and conclude with the mammals and reptiles.

The Anamnia

Amphibia

Normal development of the gonadal anlagen

Gonad formation in the urodeles and anurans can be treated together, the differences between the two groups being only of minor significance.

The genital ridges, the first visible anlagen of the gonads, arise as local thickenings of the coelomic lining, which is then called 'gonadal epithelium'. They are situated between the dorsal root of the mesentery and the Wolffian ducts and run along the ventro-lateral aspects of the mesonephric anlagen. They extend cranio-caudally from the anterior end of the mesonephros to a point beyond its posterior extremity. Only the middle portions of the genital ridges will form the definitive gonads. Here the great majority of the PGCs accumulate underneath the thickened gonadal epithelium (Wylie & Haesman, 1976). In the anurans the anterior portion of the genital ridge develops into the fat body or Bidder's organ. In the urodele gonadal anlage Stärk (1955) distinguished cord-like and band-like sections in both the pre- and post-gonadal regions.

The PGCs apparently migrate from the dorsal root of the mesentery towards the gonadal anlagen around the time of the first appearance of the ridges in the coelomic lining (Wylie & Haesman, 1976; Wylie, Bancroft & Haesman, 1976). After the arrival of the PGCs medullary tissue is formed in the genital ridges by immigration of neighbouring mesenchyme cells. In these so-called 'indifferent gonads' the PGCs are chiefly localised in the outer, cortical layer of the genital ridge, which arises from the thickened gonadal epithelium and surrounds the now solid and voluminous medullary tissue (see also Cambar *et al.*, 1970; Delbos *et al.*, 1971).

In future males the PGCs leave the cortical layer of the gonad and settle in the medullary tissue, whereas in future females they stay in the cortical layer while the medullary tissue becomes cavitated, developing into six or seven ovarial sacs (Stärk, 1955). According to Dantschakoff (1950) and Witschi (1957, 1971), among others, sexual differentiation of the potentially bisexual indifferent gonad involves the development and proliferation of one of the two components (medulla or cortex) and the simultaneous regression of the other, under the influence of antagonistic hormones which constitute the expression of the activity of the sex chromosomes. In normal development the chromosomal sex character of the germ cells (XY and XX) expresses itself in their affinity for a particular tissue environment (medulla or cortex). Under experimental conditions the sex of the surrounding somatic tissues of the gonad usually prevails over that of the germ cells, so that in chimaeric embryos sex reversal of the germ cells may occur.

The causal analysis of gonad formation

The factors responsible for gonad formation are very similar in anurans and urodeles, so that we will discuss them together, only noting some minor differences.

The first striking fact is that gonad formation is hardly affected by the absence of PGCs. This was first studied by means of UV-irradiation of the germinal plasm at the 1- to 2-cell stage of anuran eggs by Bounoure (1939), Huck & Aubry (1952), Padoa (1963a, 1964) and others. A second approach was that of extirpating blastomeres containing germinal plasm at blastula and early gastrula stages (Librera, 1964) or at the early tail-bud stage (Blackler & Fischberg, 1961). Under these experimental conditions normal genital ridges were formed (Wylie *et al.*, 1976). In urodeles removal of the presumptive lateral plate mesoderm from neurulae (Nieuwkoop, 1947) did not affect genital ridge formation either, although germ cells were entirely or almost entirely absent.

Genital ridge formation seems primarily to depend upon the presence of the endoderm. In urodeles removal of the entire endoderm from early neurulae prevented the formation of genital ridges, notwithstanding the fact that all the other neighbouring structures, including PGCs, were present (Nieuwkoop, 1947). This was confirmed for the anuran embryo by Gipouloux (1967), who found that the presence of the endodermal archenteron roof was sufficient for genital ridge formation. It must therefore be assumed that the endoderm is responsible for the evocation of the competence for genital ridge formation in the coelomic lining.

Several embryonic primordia seem to be responsible for the local appearance of the genital ridges. Removal of the notochord, the somites or the Wolffian ducts separately affects normal genital ridge formation neither in urodeles nor in anurans. In the urodeles Nieuwkoop (1947, 1950) observed accessory genital ridge formation in the coelomic lining along implanted notochord or Wolffian duct, and possibly around extra clusters of PGCs. In the anurans Cambar (1952) and Gipouloux (1967) did not observe accessory genital ridge formation along implanted notochord, somites or Wolffian ducts. It only occurred after implantation of notochord plus somites, somites plus Wolffian ducts, notochord plus Wolffian ducts, or notochord plus somites plus Wolffian ducts. This is in agreement with Cambar's and Gipouloux's observation that in anurans genital ridge formation can only be prevented when either of the following are removed: all the dorsal mesodermal primordia; notochord plus the major portion of the somites plus Wolffian ducts; notochord plus somites; or notochord plus Wolffian ducts. We must conclude that all three primordia in concert exert an influence upon the coelomic lining, previously activated by the endoderm, which leads to the local formation of the genital ridges. The competence of the coelomic lining for genital ridge formation is probably higher in the urodeles than in the anurans, since accessory genital ridge formation is apparently achieved more easily in the former than in the latter.

Ablation of the Wolffian duct prevents mesonephros formation in urodeles (Nieuwkoop, 1947; Kotani, 1962). In the anurans the effect is very similar. Houillon (1956) observed either the total suppression of mesonephros formation, or a later and rudimentary appearance of only the most anterior mesonephric anlagen in the Wolffian-duct-free region. Although Cambar (1948) did not observe mesonephros formation along a Wolffian duct grafted into non-mesonephric mesoderm, Capuron (1968) found accessory mesonephros formation along a Wolffian duct growing out ectopically in *duplicitas* embryos. Contrary to Nieuwkoop (1947), who found normal gonad formation in the mesonephros-free region obtained after unilateral Wolffian duct ablation, Houillon (1956) observed inhibition or even absence of gonad development in such mesonephros-free embryos. The inhibitory effect was abolished, however, when the contralateral gonadal anlage was also removed. This suggests that a genital ridge was initially formed but was affected only later by the absence of the mesonephros.

In anurans ablation of the Wolffian duct does not initially affect the normal cranio-caudal distribution of the PGCs, but later the local absence of the mesonephros seems to lead to a marked degeneration of PGCs, which persist only near remaining mesonephric anlagen, where local gonad development occurs (Houillon, 1956). The Wolffian duct moreover seems to be indispensable for the normal elongation of the genital ridges at feeding stages.

The absence of a mesonephric blastema, caused by ablation of the Wolffian duct, affects neither the formation of the inter-renal blastemata nor that of the medullary tissue in the indifferent gonad (Cambar & Mesnage, 1963). This observation was confirmed by Vannini & Giorgi (1969). Apparently the medullary tissue (in the form of medullary cords) of the gonad is not derived from the mesonephric anlagen but from inter-renal mesenchyme.

Absence of medullary tissue leads to an inhibition of PGC multiplication, both in males and females, while excessive medullary tissue causes an abnormally rapid proliferation of the PGCs (Gipouloux, 1973). Therefore, the mutual cooperation of cortex and medulla as well as the presence of PGCs is required for normal gonad development.

Vannini (1962) found in *Bufo* that the regional differentiation of the genital ridges into Bidder's organ and definitive gonad is not affected by the absence of the mesonephros. Whereas the gonadal region can form Bidder's organ by regulation after removal of the anterior part of the genital ridge, regulation does not occur after removal of the posterior part. The reader is further referred to Cambar (1948), Vannini & Gardenghi (1964), Blackler (1966) and Gipouloux (1973).

In the Coecilia a very primitive gonad is found. It has a typical segmental character corresponding topographically to the muscle segments. The segmental structure is maintained in the developing testis but becomes rather obscure in the developing ovary. The fat body, which runs parallel to the gonad, is likewise segmental in structure, both in males and females, demonstrating the intrinsic segmental character of the gonadal anlagen (Wake, 1968).

Fishes

The gonadal anlagen in the Agnatha are typically hermaphroditic. They extend about half the length of the animal and contain two kinds of germ cells, small ones showing rapid proliferation and large ones showing a tendency to further growth. In the lampreys sex determination occurs as a result of predominance of one of the two germ cell types (Okkelberg, 1921), genetic males showing a predominance of the former type, genetic females of the latter. In *Myxine* and *Bdellostoma* the large type of germ cells are found particularly in the cranial region of the gonad and the small type in the caudal region. Here sex differentiation is achieved by regression of one of the two regions (Schreiner, 1955; see also Franchi *et al.*, 1962).

In the Chondrichthyes there are distinct cortical and medullary components in the gonad. Sex differentiation occurs as in the amphibians by the proliferation of one component and the regression of the other (Chieffi, 1959; see also Franchi *et al.*, 1962).

Hardly anything is known about gonad development in the Osteichthyes. De Smet (1970) described the indifferent gonad of *Polypterus* as having a mesenchymal core, which is rich in melanophores, surrounded by cuboidal epithelium. The definitive gonad develops in the region of the eighth to sixteenth somites, where it is colonised by PGCs.

Little is known about gonad formation in the Teleostomi. The PGCs accumulate underneath the gonadal epithelium prior to actual gonad formation (Johnston, 1951; Belsare, 1966). Gamo (1961*b*) states that the mesonephros does not participate in gonad formation. Moreover there is no clear distinction between cortex and medulla. Sex differentiation occurs in the female by the enlargement of the germ cells (oogenesis) and the subsequent cavitation of the gonad, while in the male spermatogenesis and testicular tubule formation begin at a later stage of development (D'Ancona, 1950; Johnston, 1951; Belsare, 1966). According to Johnston (1951) the gonad develops only where PGCs are present. (See also the review of Vivien, 1964.)

The lower chordates

Little is known about gonad formation in these groups. For the Cephalo-chordata there is the original description of the development of the strictly segmental gonadal anlagen in *Branchiostoma* by Boveri (1892) (fig. 6.1). Gonads are formed at the level of the eleventh to thirty-sixth somites by an outpocketing of the myocoel on the ventro-cranial side of each somite. This outpocketing engulfs the segmentally located PGCs. Later the gonadal pouch separates from the rest of the myocoel and subsequently the connection of the cluster of PGCs with the myotome is broken. When the individual gonadal pouches enlarge they may ultimately touch each other, so that the gonad takes the form of a string of pearls (Wickstead, 1976).

In the Tunicata a hermaphroditic gonad is formed around a group of haemoblasts (cf. chapter 5, p. 96). It is either an ovotestis, when an oocyte is present, or a testis. The ovarian and the testicular portion of the hermaphroditic gonad are connected to separate diverticula of the atrial cavity (Mukai & Watanabe, 1976). (See further the reviews by Reverberi, 1971*a*, and Berrill, 1975.)

The Amniota

Aves

Our knowledge of gonad formation in birds is for the greater part descriptive. The few experimental data will therefore not be treated separately but together with the descriptive data.

In seeming contrast to the lower vertebrates the gonadal anlagen are found first in the dorsal splanchnic mesoderm on both sides of the dorsal mesentery as local thickenings of the coelomic epithelium. They subsequently shift through the coelomic angles towards the dorsal somatic mesoderm, where they are ultimately found near the mesonephric anlagen. Since the morpho-genetic stimuli responsible for the local formation of the gonadal anlagen in birds are unknown (Dubois & Croisille, 1970) not much can be said about their divergent site of origin. It may be the consequence of differential growth of the two layers of the lateral plate mesoderm prior to the appearance of the gonadal anlagen. The two gonadal anlagen are initially of equal size but

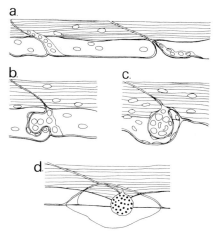

Fig. 6.1. Stages in the formation of segmental gonadal anlagen around PGCs at ventro-cranial extremities of somites in *Branchiostoma lanceolatum* (compare Fig. 5.3, p. 97). (After Boveri, 1892.)

develop differently, the left/right (L/R) inequality becoming very pronounced in the ovarian anlagen but remaining only slight in the testicular ones (Blocker, 1933; Van Limborgh, 1957, 1958). Rete cord and medullary tissue, which forms between stages 23 and 26 (H. & H.), is of extra-gonadal origin (Witschi, 1951; Van Limborgh, 1957, 1958) and may derive from the mesonephric anlagen (see also Franchi *et al.*, 1962), since a deficient mesonephros leads to subnormal gonadal development (Didier & Fargeix, 1976a).

The first proliferation of the gonadal epithelium, which takes place in the 4- to 5½-day embryo, leads to the formation of the primary sex cords of the indifferent gonad. The R ovary shows no further development but begins to regress at day 8 and reaches its minimal size 7 to 10 days after hatching. The L ovary shows a second proliferation of the gonadal epithelium at day 6, leading to the formation of the definitive ovarian cortex, in which meiosis starts at day 11. A second proliferation of the gonadal epithelium is lacking or very transitory in the testis, where the primary sex cords form the seminiferous tubules (e.g. Franchi *et al.*, 1962; Kannankeril & Domm, 1968). Although the amount of medullary tissue differs only slightly between the L and R gonads, both in male and female, the amount of cortical tissue in the L ovary exceeds by many times that in the R ovary.

PGCs begin to colonise the gonadal anlagen between stages 16 and 19 (H. & H.). The initial colonisation of the two anlagen is essentially equal (Van Limborgh, 1957; 1958; Simon, 1960; Didier & Fargeix, 1976a, b) but later the number of PGCs in each is different. This is not due to a redistribution of the PGCs between the two gonads (Van Limborgh, 1957, 1958; cf. Witschi, 1935) but to differential changes in the numbers of PGCs in the two anlagen. As a consequence the onset of sexual differentiation is characterised by a markedly unequal number of PGCs in the two gonads in the female embryo. Van Limborgh (1960, 1961) initially ascribed this asymmetry exclusively to a much more pronounced degeneration of PGCs in the R than in the L gonad. In 1968 he stated, however, that the subsequent increase in the number of PGCs between stages 22 and 27 (H. & H.) is predominantly in the PGCs in the cortex of the L ovary. The number of PGCs in the R ovarian cortex is nearly equal to that in the L and R testicular cortex, while the number of PGCs found in the medullary tissue is the same in male and female gonads.

Contrary to Dantschakoff's (1932b) original statement, the absence of PGCs prevents neither normal gonad formation nor subsequent sexual differentiation (Willier, 1937). The further development of the sterile ovary, however, is arrested after the second proliferation of the gonadal epithelium, and that of the sterile testis after the formation of the primary sex cords (Dulbecco, 1946, 1948). It is evident that gonadal asymmetry must be based on factors intrinsic to the gonadal anlagen.

Fargeix (1967) found a pronounced discrepancy between the number of PGCs in the germinal crescent and that in the gonadal anlagen, due to the limited capacity of the gonadal anlagen to accommodate PGCs. This conclusion was supported by the observation of Didier & Fargeix (1974) and Didier, Fargeix & Bergeaud (1974) that after the early removal of one gonadal anlage no excess of PGCs was found in the remaining one. The L/R asymmetry of the gonadal anlagen seems to be based on an intrinsic asymmetry of the early blastoderm, since L and R twin embryos resulting from bisection of unincubated blastoderms have after 6 days of incubation, numbers of PGCs which correspond to those found in the L and R gonadal anlagen of a normal embryo (Fargeix, 1970).

Mammalia

Gonad development in the mammals is very similar to that in birds. Sexual differentiation likewise seems to be based on an antagonism between the development of the gonad's two components, cortex and medulla. The cortex constitutes the dominant structure in the female, the medulla in the male (see Franchi *et al.*, 1962). Pelliniemi (1975) studied the ultrastructure of the gonadal ridge in the pig embryo at 21 days of gestation and found that the primary sex cord cells have the same ultrastructure as the gonadal epithelium. In the 38- to 39-day human embryo Pelliniemi (1976) described a stratified gonadal epithelium with a discontinuous basal lamina.

Gaillard (1950) found that the presence of gonadal epithelium is vital to the survival of explants of ovarian tissue. A stable situation with subsequent differentiation is only reached when the gonadal epithelium surrounds the entire explant. In such explants Gaillard (1952) observed the formation of new parenchymal cords.

As in the other vertebrates, the presence of PGCs is not necessary for normal gonadal development. Everett (1943) found normal proliferation of the gonadal epithelium in grafts of genital ridges taken before PGC colonisation and implanted into the kidney capsule of host animals. Normal gonad development was also obtained after sterilisation of the embryo with X-rays. This was confirmed by several authors, most recently by Merchant-Laros (1976), who observed a normal formation of the indifferent gonad followed by normal morphogenesis of the foetal ovary in bisulphan-sterilised rats. He found only indications that the presence of PGCs may be necessary for normal endocrine activity of the gonad.

In her 1970 review Peters states that the first indication of sex differentiation in the 12-day male mouse embryo is the formation of seminiferous cords, while its first symptom in the female is the appearance of meiotic prophases in the germ cells (Simkins, 1923; Brambell, 1927). However, Mintz (1960*b*) observed the first signs of sexual differentiation in mouse embryos to be the distribution of the PGCs at 11 days. In the female embryo the PGCs maintain a peripheral position, which is followed by pronounced cortical development, whereas in the male embryo the PGCs acquire a more central position, coinciding with a predominance of the medullary component of the gonad.

Reptilia

Very little is known about gonad formation and development in the reptiles. According to Mendietta (1963) gonad development in *Gongylus* (Lacertilia) occurs in three steps, viz. (1) formation of the undifferentiated gonadal anlage, (2) formation of a potentially bisexual, indifferent gonad, and (3) differentiation into ovary or testis. In the indifferent gonad no clear subdivision into cortex and medulla is yet to be seen. A segregation of the two components occurs at a more advanced stage of development. In the reptiles the formation of the ovary also seems to be characterised by a further development of the cortex and a regression of the medullary component, while in the development of the testis the reverse situation obtains. (See also Simkins & Asana, 1931.)

General conclusions

It seems likely that in the chordates the most primitive state of gonad formation is represented by the segmental gonadal anlagen found in the Cephalochordata. In the vertebrates a segmental arrangement of the gonadal anlagen is found only in the Coecilia, the most primitive amphibia.

It seems probable that gonad formation is governed by the same principles in all vertebrates, as regards the formation of the indifferent gonad as well as its differentiation into either ovary or testis. In the Anamnia the local formation of the indifferent gonad has been most thoroughly investigated in the amphibians, where after an initial sensitisation of the coelomic lining by the endoderm (appearance of competence) a combined inductive action of notochord, Wolffian duct and probably somites leads to the local appearance of the genital ridges. In the subsequent formation of the medullary tissue the inter-renal mesenchyme seems to play a major role, while the mesonephros may only be of importance for the maintenance of the gonadal anlagen. Indifferent gonad formation in the Amniota differs from that in the Anamnia by the additional formation of primary (and secondary) sex cords, representing the first (and second) proliferation(s) of the gonadal epithelium. In the amniotes the causal factors responsible for gonad formation are unfortunately unknown. A fact which particularly requires further explanation is that in birds the initial position of the gonadal epithelium is in the dorsal splanchnic mesoderm, whereas in the other vertebrates its first appearance is in the dorsal somatic mesoderm.

7

The migration of the primordial germ cells

With the exception of the Cephalochordata, where the PGCs arise *in situ* in their definitive location and no migration is therefore required, in all the vertebrate groups a displacement of the PGCs occurs from their extra-gonadal (in the amniotes extra-embryonic) site of initial appearance to their ultimate destination in the gonadal anlagen.

The migration of the PGCs from their site of origin towards the gonadal anlagen has been most thoroughly studied in birds, and reasonably well studied in the anuran amphibians. Since the migration route is quite different in the two groups, and since as far as we know these groups essentially typify the two main routes along which PGCs may reach the gonadal anlagen, we will first focus our attention on the anuran amphibians, deal only briefly with the urodeles and fishes, subsequently discuss extensively the situation in the birds, and finish with some additional data on the mammals and reptiles.

It seems desirable to distinguish between two different kinds of movement of the PGCs during their migration, i.e. (*a*) 'passive' displacement by morphogenetic movements, and (*b*) 'active' displacement by their own amoeboid motility. During their active displacement the PGCs may make use of the circulatory system of the embryo. This is often called 'passive transport' but should be clearly distinguished from the passive displacement by morphogenetic movements, during which the PGCs do not show any motility. The term 'vascular transfer' therefore seems more appropriate.

The Anamnia

Amphibia

Anura

The displacements of the presumptive and true PGCs in the anuran amphibians have already been described in the preceding chapters.

It is now generally agreed that the displacement of the germinal plasm from the vegetal surface of the egg towards a more internal position during cleavage is due to the morphogenetic movements involved in the cleavage process. The subsequent displacements of blastomeres containing germinal plasm (pPGCs) towards the centre of the yolk mass and even as far as the floor of the

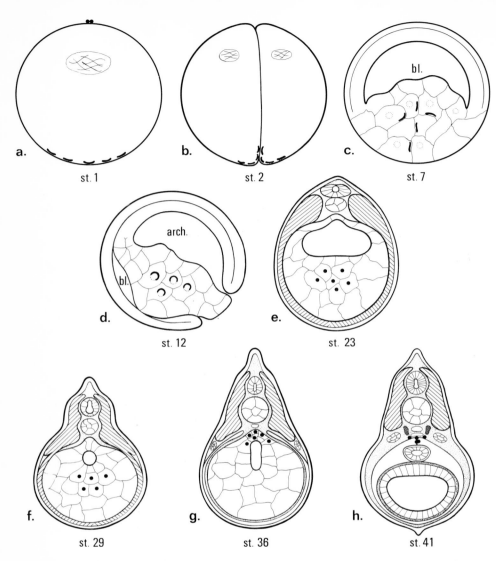

Fig. 7.1. Fate of germinal plasm and migration of PGCs in the anuran amphibians. (a) Localisation of germinal plasm in the form of subcortical cytoplasmic patches near the vegetal pole of the fertilised, uncleaved egg. (b) As for (a), but at the 2-cell stage. Beginning of ascent along cleavage furrow. (c) Internal (passive) displacement of endodermal blastomeres containing germinal plasm by pregastrulation movements at early blastula stage. (d) Further (passive) displacement of endodermal blastomeres by gastrulation movements, and intra-cellular displacement of germinal plasm from a peripheral towards a juxta-nuclear position. (e) and (f) Nearly unchanged localisation of PGCs at tail-bud stages. (g) Active dorsal migration of PGCs inside endoderm at early larval stage and accumulation in dorsal endodermal ridge. (h) Further active migration of PGCs through dorsal mesentery towards genital ridges. arch., archenteron; bl., blastocoel.

blastocoel (Bounoure, 1939) can likewise be satisfactorily explained by the morphogenetic pregastrulation movements, which were first described by Schechtmann (1934) in *Triturus* and analysed further by Gipouloux (1962) in *Discoglossus*. The displacements of the PGCs during gastrulation and neurulation also seem to be due to morphogenetic movements affecting the endodermal yolk mass. Since up to this stage of development no active movements of the PGCs are required, these displacements of the PGCs and their forerunners may be designated as *passive displacement*. (fig. 7.1a–d). (See for example Bounoure, 1939; Blackler, 1958, 1966; Whitington & Dixon, 1975.)

At the end of this first period of development the PGCs are located in the centre of the yolk endoderm in the caudal half of the embryo. Kamimura *et al.* (1976) found that in *Xenopus* the PGCs subsequently move to a peripheral position in the yolk mass (stage 31, N. & F.). They then move to the dorsal endoderm during stages 33 to 36, accumulate in the form of a dorsal endodermal crest or ridge around stage 40, and separate from the endoderm at stage 41. Subsequently they move to the dorsal root of the mesentery, and finally laterally towards the genital ridges (fig. 7.1e–h). Wylie *et al.* (1976) observed that in *Xenopus* the PGCs actually accumulate underneath the gonadal epithelium of the genital ridges.

Vannini & Giorgi (1969) and Giorgi (1974) hold the opinion that the displacements of the PGCs from the dorsal endodermal crest towards the dorsal root of the mesentery and subsequently towards the genital ridges are mainly caused by growth processes in the surrounding tissues (passive displacement). There is, however, rather convincing evidence that the displacements of the PGCs during this phase of development are for the greater part due to their active movements.

As already mentioned in chapter 3 (p. 62) the PGCs show a distinct cell boundary in the light microscope. When studied with the electron microscope this is found to be due to wide intercellular spaces between the PGCs and the surrounding cells. Cambar *et al.* (1970) described the ultrastructure of the PGCs in *Rana* during their migration and after they have settled in the genital ridges. The PGCs show cytoplasmic processes extending into the intercellular spaces (Kamimura *et al.*, 1976). With time-lapse cinematography Wylie & Roos (1976) actually observed active amoeboid movements *in vitro* of PGCs isolated during the migration period.

Gipouloux (1967) carried out an experimental analysis of the displacements of the PGCs in *Discoglossus* and other anurans. After dorso-ventral transposition of the endodermal mass the PGCs migrate through the endoderm in a direction opposite to normal, though in reduced numbers. In non-rotated endoderm there was normal dorsal migration (PGCs in dorsal root of mesentery) when either notochord plus somites plus Wolffian ducts, notochord plus somites, or somites plus Wolffian ducts were present. Partial migration (PGCs partially in dorsal endoderm and partially in mesentery) was observed when somites alone were present, and still more restricted migration (PGCs in dorsal endoderm only) in the presence of part of the somites or of the Wolffian ducts only. Implanted notochord or Wolffian duct exerted a weak attraction on only some of the host PGCs. The effect was stronger with implanted somites, stronger still with implanted notochord plus Wolffian ducts or with notochord plus somites, and strongest with implanted notochord plus somites plus Wolffian ducts. Gipouloux concludes from these experiments that the dorsal mesodermal organs exert an attracting influence upon the PGCs, leading to the establishment of a dorso-ventral gradient of a chemotactic agent inside the embryo. This primary attraction of PGCs is a property of the caudal, but not of the cephalic mesoderm (Giorgi, 1974). In embryos brought into ventral parabiosis, so that two axial systems are situated in opposite sides of a single endodermal mass, the PGCs are partially immobilised by the opposite attracting influences, leading to a markedly reduced migration.

That the attractant is a chemical agent is rendered likely by the observation that PGCs are also attracted by an implanted piece of agar previously immersed in an extract of dorsal mesodermal structures (Gipouloux, 1967). However, Gipouloux's study of the diffusibility of the attractant through millipore filters and of its heat and enzymatic degradation is in our opinion too preliminary to allow any definite conclusion regarding the possible chemical nature of the attractant.

Gipouloux (1967) also found that suppression of genital ridge formation by removal of the axial structures or by mercuric chloride treatment leads to a 'permanent' accumulation of the PGCs in the dorsal root of the mesentery. Nieuwkoop (1947) had observed an attraction of host PGCs by an extra genital ridge formed in urodele larvae after implantation of either notochord or Wolffian duct anlagen. This was confirmed by Gipouloux (1967) for the anuran embryo. Moreover, there is the interesting observation by Martin (1959) that in *Alytes obstetricans*, which has L and R genital ridges of unequal lengths, the PGCs are also distributed unequally between the two gonadal anlagen. These data plead strongly in favour of a chemotactic action exerted by the genital ridges.

Urodela

Nieuwkoop (1947) concluded for his xenoplastic transplantations of ventro-lateral marginal zone (see chapter 5, p. 88) that in the urodeles the PGCs are initially more or less evenly distributed throughout the presumptive lateral plate mesoderm. This was confirmed by Maufroid & Capuron (1973) by means of grafts of ventral and ventro-lateral marginal zone into the

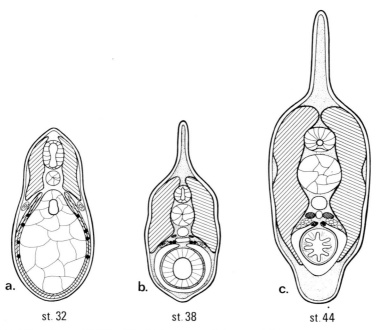

Fig. 7.2. Migration of PGCs (black) in the urodelen amphibians. (a) Successive steps in the migration of PGCs along the mesoendodermal interspace towards the dorsal midline, and (b and c) their subsequent migration towards the genital ridges.

blastocoelic cavity of host embryos (see chapter 5, p. 89). Maufroid & Capuron (1972) studied the dorsal displacements of the PGCs in *Pleurodeles* during stages 16 to 27 (Gallien & Durocher, 1957) by means of unilateral ablation of various dorso-ventral regions of the lateral plate mesoderm (compare the schematic fig. 7.2). Ikenishi & Nieuwkoop (1978) found cilia on some of the PGCs in *Ambystoma* during the migration period. However, nothing is known about the mechanisms upon which the displacement of the PGCs through the lateral plate mesoderm and their ultimate accumulation in the genital ridges are based.

Coecilia

The migration of the PGCs in the Coecilia has not been studied.

Fishes

The only evidence available on the migration of the PGCs in the fishes is of a descriptive nature.

In the lamprey (Agnatha), where according to Okkelberg (1921) the PGCs seem to originate in the lateral plate mesoderm, the migration route resembles that in the urodele amphibians.

Among the Osteichthyes Allen (1911) described the migration route in *Amia* and *Lepisosteus*. In *Amia* the PGCs migrate from the peripheral endoderm into the lateral plate mesoderm, then accumulate in its mediodorsal region, and migrate from there to the genital ridges. This situation seems more or less intermediate between that in the anuran and that in the urodele amphibians. In *Lepisosteus* the PGCs first migrate towards the dorsal portion of the gut and subsequently through the dorsal mesentery to the genital ridges, a route which resembles that in the anuran amphibians.

No relevant data are available on the migration of the PGCs in the Chondrichthyes.

In the Teleostomi the available data are scanty and controversial. The authors who assume an endodermal origin for the PGCs take them to migrate from the yolk sac endoderm – or even earlier from the periblast – through the splanchnic mesoderm towards the dorsal midline, and subsequently through the dorsal mesentery towards the genital ridges (Richards & Thompson, 1921, in *Fundulus*; Johnston, 1951, in *Micropterus*; Belsare, 1966, in *Channa*; Hogan, 1973, in *Oryzias*). The authors who defend a mesodermal origin of the PGCs assume them to arise in the unsegregated mesendoderm. After segregation of the two layers they are found in the mesodermal layer or distributed throughout all three germ layers, with a preference for the mesodermal layer (Gamo, 1961*b*, in *Oryzias*). Although the PGCs may show a temporal association with the peripheral endoderm, they subsequently migrate through the splanchnic towards the somatic mesoderm and further towards the genital ridges (Gamo, 1961*c*, in *Oryzias*; Pala, 1970, in *Gambusia*; Nedelea & Steopoe, 1970, in *Cyprinus*).

Several authors have observed pseudopod formation on the surface of the PGCs, which they took to be an indication of active migratory movements (Johnston, 1951; Nedelea & Steopoe, 1970), but other authors were more inclined to ascribe the movements of the PGCs entirely (Richards & Thompson, 1921, in *Fundulus*) or at least partially (Hogan, 1973, in *Oryzias*) to passive displacement by the surrounding tissues. Stolk (1958), who observed many highly ectopic PGCs in *Abramites* and *Cirrhina*, even assumed vascular transfer to be involved, but this has never been confirmed.

The Amniota

Aves

The presumptive PGCs are very probably passively displaced by the morphogenetic movements occurring during the formation of the primary and secondary hypoblast and the subsequent formation of the primitive streak and

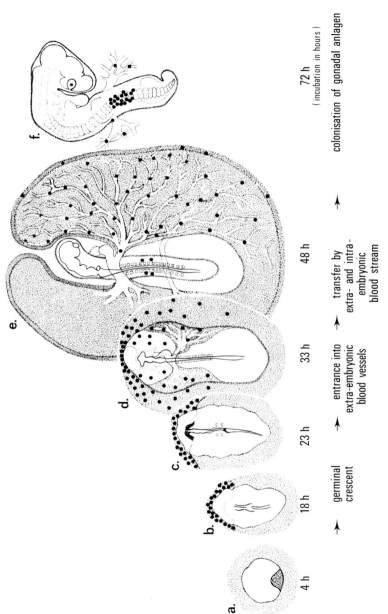

Fig. 7.3. Migration of PGCs in the avian embryo. (a) Absence of identifiable germ cells prior to primitive streak formation. (b) and (c) (Passive) accumulation of PGCs in the anterior germinal crescent at head-process and head-fold stages, respectively. (d) (Active) penetration of PGCs into mesodermal blood islands and beginning of transfer by blood circulation. (e) Circulation of PGCs through entire vascular system and start of their egress from the vascular system in the neighbourhood of the gonadal anlagen. (f) Continuing colonisation of gonadal anlagen by actively migrating PGCs.

the invagination of the tertiary hypoblast or embryonic endoderm (Dubois, 1967) (see also chapter 2, p. 45 and chapter 5, p. 97). The subsequent, active displacements of the PGCs from the germinal crescent towards the genital ridges may be divided into four phases. The first phase is their segregation from the endodermal layer and their accumulation between the endoderm and ectoderm during stages 4 to 8 (H. & H.). In the second phase the PGCs begin to penetrate the vascular network. This occurs after the invagination of the extra-embyronic mesoderm, its subsequent extension between the ectoderm and endoderm, and the start of the differentiation of the blood islands (stage 10, H. & H.). In the next phase the PGCs are found successively in the extra-embryonic blood vessels at stage 12, and in the embryo proper with the onset of cardiac pulsation and blood circulation at stage 13. By $2\frac{1}{2}$ days of incubation (last phase) they start to leave the visceral branches of the aorta – the gonadal anlagen at that time being situated close to the omphalo-mesenteric artery – and begin to penetrate the gonadal epithelium. The majority of the PGCs have settled in the gonadal epithelium by 3 days of incubation (fig. 7.3).

During the formation of the indifferent gonads the PGCs are scattered in the cortical layer but also in the medullary tissue. In the young ovary they accumulate in the cortex formed by the second proliferation of the gonadal epithelium, but migrate into the medullary tissue in the young testis (e.g. Blocker, 1933; Meyer, 1964; Fujimoto *et al.*, 1976a).

Simon (1957) translocated the anterior half of the blastoderm containing the germinal crescent laterally with respect to the posterior half containing the presumptive gonadal anlagen, and in this way showed that the PGCs actually reach the gonadal anlagen by the only plausible route, i.e. the vascular system. In 1960 Simon provided an even more elegant demonstration of the vascular transfer of the PGCs by joining a blastoderm sterilised by removal of the germinal crescent to a normal one by parabiosis. Through the anastomosing vascular systems the gonads of the two embryos were colonised with PGCs. In 1961 she performed parabiosis between an anterior half embryo containing PGCs and a posterior half embryo containing gonadal anlagen, and obtained normal colonisation of the gonadal anlagen with PGCs through the connecting vascular system. (See also Simon's review of 1964.)

Dantschakoff's (1936) notion that the PGCs would leave the vascular system by being mechanically trapped in the fine capillaries of the gonadal region – an idea also adhered to by Blocker (1933) – was disposed of by Van Limborgh, Van Deth & Tacoma (1960), who found that the capillaries of the head and the yolk sac are in fact finer and thus more suitable for trapping PGCs than those of the gonadal region; nevertheless the PGCs only leave the vascular system in the vicinity of the gonadal anlagen. Van Limborgh *et al.* proposed instead that a slowing-down of the blood stream in the capillary network promotes the egress of the PGCs from the vascular system. Simon (1960) came to the same conclusion and suggested that in addition a chemotactic process is responsible for the directive migration of the PGCs (see below). The former idea was confirmed by Dubois (1968), who observed that an experimental reduction of the blood flow causes a greater number of PGCs to leave the large blood vessels. The discovery of PGCs in blood smears of 2-day embryos, using both phase contrast microscopy and PAS staining, demonstrated unequivocally that the PGCs do indeed use the vascular route to reach the gonadal anlagen (Singh & Meyer, 1967; Fujimoto *et al.* 1976*b*). Further support for this route is given by Reynaud (1969), who was able to repopulate the gonads of embryos previously sterilised by UV-irradiation of the germinal crescent by intravenous injection of a cell suspension made from germinal crescent endoderm of another embryo.

The insertion of a piece of shell membrane between the hypoblast and the presumptive vascular mesoderm does not prevent the PGCs from leaving the hypoblast (although it of course prevents their entering the vascular system), and an isolated hypoblast cultured *in vitro* shows a spontaneous but random migration of the PGCs (Dubois, 1969). Thus, the first step in the migration of the PGCs, i.e. their movement from the endodermal layer of the germinal crescent into the vascular system, is an active process. PGCs not only have amoeboid motility but actively invade various tissues under in-vitro conditions. The entry of the PGCs into the vascular system apparently does not require a specific attraction by the vascular mesoderm.

As already suggested by Simon (1960) the PGCs leave the vascular system under the influence of a chemotactic attraction exerted by the gonadal epithelium. *In vitro* the PGCs may leave the germinal crescent when it is associated with the gonadal region of another embryo, and may colonise the gonadal epithelium in the absence of a vascular system (Dubois, 1968). The attraction exerted by young gonadal epithelium can also act upon the PGCs in a colonised but still undifferentiated gonad. It turned out that even spermatogonia still possess migratory capacity, sensitivity to the attractive influence, and the ability to colonise gonadal epithelium at least up to the 12th day of incubation. The majority of the oogonia lose their migratory capacity on the 8th day. Since the attraction can be exerted across a permeable barrier such as the shell membrane, the most satisfactory explanation is that of positive, selective chemotaxis.

By delayed injection of a PGC suspension made from germinal crescent endoderm, Reynaud (1969) was able to show that a sterile host gonad obtained by UV-irradiation of the germinal crescent can still trap PGCs at 5 days of incubation. Thus the competence of the PGCs to react to the attracting influence lasts much longer than the normal period of PGC immigration and only disappears with the onset of gametogenesis, while the attracting capacity of the gonad disappears during the formation of the primary sex cords (Reynaud, 1971*b*). Rogulska (1969) demonstrated a direct transfer of PGCs from an intracoelomic graft of germinal crescent to an adjacent host genital ridge at $3\frac{3}{4}$ days of incubation, i.e. long after normal migration is completed. By means of intracoelomic grafts of mouse hindgut to the gonadal region of a chick embryo, Rogulska, Ozdzenski & Komar (1971) found that chick gonadal epithelium also attracts mouse PGCs, demonstrating that the attractant is not species or class specific.

Let us now consider the properties of the PGCs and the gonadal epithelium more closely in order to understand better the chemotactic process. As already mentioned under the characteristics of the PGCs listed in chapter 3 (p. 64), the PGCs in birds undergo a measure of cytogenesis. Clawson & Domm (1963*a*, *b*) described the PGCs of the germinal crescent as containing much yolk and only little glycogen. The migrating PGCs, however, contain far less yolk but much more glycogen. The glycogen content diminishes during migration, so that after arrival in the gonadal ridges the PGCs again contain little glycogen. This was confirmed by Fujimoto *et al.* (1975). Dubois & Cuminge (1968) described more or less parallel changes in the amount of osmophilic lipid droplets, an observation confirmed by Ukeshima & Fujimoto (1975). In addition, Fujimoto *et al.* (1976*a*) described a structural change in the glycogen and its unipolar intracellular localisation during the settling of the PGCs in the gonadal epithelium. The glycogen and the lipids are considered to play an important role as energy sources during the migration of the PGCs (Dubois & Cuminge, 1968).

The PGCs show pseudopodial extensions during their migration (Ukeshima & Fujimoto, 1975; Fujimoto *et al.*, 1967*a*, *b*). Microfilaments and microtubules, however, have not been demonstrated (Dubois & Cuminge, 1968). According to Cuminge & Dubois (1971) the PGCs show an intense lytic activity and actively penetrate the gonadal epithelium. There they take up soluble substances and cellular debris by pinocytosis and phagocytosis, respectively (see review by Dubois & Croisille, 1970).

The gonadal epithelium has a syncytial and spongy character and shows large intercellular spaces. The same features are found in sterile gonadal epithelia, so they are not due to the lytic activity of the PGCs. Autoradiographic electron-microscopic analysis has shown that the gonadal epithelium synthesises exportable proteins or protein complexes, which are set free by a merocrine excretory process (Cuminge & Dubois, 1971). Although in the nucleus only a single slow type of protein replacement is observed, two classes of proteins are synthesised in the cytoplasm, i.e. slowly replaced structural proteins and rapidly replaced exportable glycoproteins (Cuminge & Dubois, 1971). However, it has not been conclusively shown that the attractive factor produced by the gonadal epithelium is in fact a protein (Dubois & Cuminge, 1970). Swartz (1975) obtained some evidence that the chemical attractant may be steroid in nature, since injection of androgens or oestrogens interferes with PGC migration. In the synthesis of the exportable proteins the Golgi complex plays an important role (Cuminge & Dubois, 1971). The fact that concanavalin A, which binds to glycoproteins, inhibits the migration of PGCs placed under the attracting influence of young gonadal epithelium, suggests that glyco-proteins play a decisive role in the chemotactic migratory capacity of the PGCs (Dubois & Cuminge, 1975).

In contrast to those of the gonadal epithelium, the glycoproteins synthesised by the PGCs show only a slow turnover and are not excreted. They are found preferentially in the lysosomal apparatus and on the plasma membrane, where they are concentrated on microvilli, pseudopodia and at sites of pinocytosis and phagocytosis (Dubois & Cuminge, 1975). (See also reviews by Cuminge & Dubois, 1971; and Dubois & Cuminge, 1974.)

Very little is known about possible attractive influences that play a role in the ultimate distribution of the PGCs between the cortical and medullary tissues of the differentiating gonad.

Summarising, it may be said that the initial step in the migration of the PGCs, i.e. their movement from the hypoblast of the germinal crescent into the vascular system, is an active but non-directional process. Impressive evidence argues for the involvement of a chemotactic mechanism in the migration of the PGCs from the vascular system into the gonadal epithelium. Very little is known about the mechanisms involved in the ultimate dis-placements of the germ cells inside the differentiating gonad.

Mammalia

Leaving aside their possible primary site of origin, the PGCs are clearly recognisable for the first time in presomite and early somite stages in the yolk sac endoderm and adjacent splanchnic mesoderm near the allantoic evagination (fig. 7.4b) (Vanneman, 1917, in the marsupial *Tatusia*; Witschi, 1948, in man; Chiquoine, 1954, in the mouse; Gomes-Ferreira, 1956, 1957, in the guinea pig; Chrétien, 1966, in the rabbit). Slightly later they are found in the endoderm and adjacent mesoderm of the hindgut. They then leave the gut and migrate through the splanchnic mesoderm towards the dorsal mesentery, and from there towards the genital ridges (fig. 7.4c, d) (Vanneman, 1917, in *Tatusia*; Celestino da Costa, 1937, and Gomes-Ferreira, 1957, in the guinea pig; Everett, 1943, Chiquoine, 1954, Merchant & Zamboni, 1973, Zamboni & Merchant, 1973, and Spiegelman & Bennett, 1973, in the mouse; Gomes-Ferreira, 1956, and Chrétien, 1966, in the rabbit; Witschi, 1948, Falin, 1969, Fuyuta *et al.*, 1974, and Fujimoto *et al.*, 1977, in man). It is still unclear whether the PGCs migrate through the endodermal epithelium of the yolk sac and hindgut, or through the adjacent splanchnic mesoderm, associating themselves temporarily with the endodermal cells for nutritive purposes only

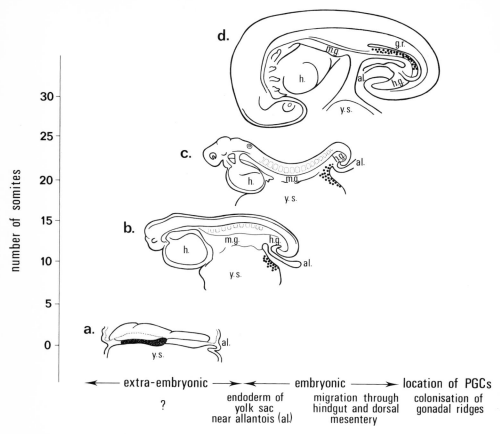

Fig. 7.4. Migration of PGCs in the mammalian embryo. (a) Absence of identifiable germ cells at the neural-plate stage. (b) Appearance of PGCs in the wall of the yolk sac. (c) (Passive) displacements of the PGCs during hindgut formation. (d) Colonisation of the gonadal anlagen by actively migrating PGCs. al., allantois; g.r., gonadal ridge; h., heart; h.g., hindgut; m.g., midgut; y.s., yolk sac.

(see Zamboni & Merchant, 1973; Spiegelman & Bennett, 1973; Clark & Eddy, 1975).

The great majority of authors hold the opinion that the initial displacements of the PGCs during gut formation are mainly or entirely passive, i.e. they result from the morphogenetic movements associated with gut formation. Opinions differ as to the mode of their subsequent translocation from the hindgut to the genital ridges. Some authors have observed large intercellular spaces around the PGCs and the formation of pseudopodia containing microtubules, which would suggest active migration (Blandau *et al.*, 1963, in squash preparations of mouse mesentery and hindgut; Jeon & Kennedy, 1973, Merchant & Zamboni, 1973, Zamboni & Merchant, 1973, Spiegelman & Bennett, 1973, in the mouse; and Falin, 1969, and Fujimoto *et al.*, 1977, in man). On the other hand, Jeon & Kennedy (1973) found tight junctions between PGCs and adjacent somatic cells, which pleads against active migration, but Spiegelman & Bennett (1973) deny the presence of intercellular junctions. Nearly all authors who favour active migration of the PGCs in the mammals assume interstitial migration through the various tissues. Semenova-Tian-Shanskaya (1969) in addition advocates partial transfer by way of the vascular system. Witschi postulated as early as 1948 a chemotactic mechanism for PGC migration, but relevant experimental evidence is lacking. McKay *et al.* (1953), Chiquoine (1954), Mulnard (1955) and McAlpine (1955) found a positive alkaline phosphatase reaction not only in the PGCs but also in the gonadal epithelium. (See also the reviews by Mintz, 1960*a*, and Peters, 1970.)

Gondos & Hobel (1971) found indications of active migration of the germ cells towards the periphery of the seminiferous tubules in the male gonad, in the form of pseudopodial extensions and the presence of microtubules. Blandau *et al.* (1963) observed undulating surface movements of oogonia in ovarial squash preparations.

In summary, it may be said that in the mammals most of the evidence pleads in favour of active, interstitial migration of the PGCs from the hindgut towards the genital ridges, whereas the earlier displacements from the yolk sac to the hindgut are probably mainly passive. The route taken by the PGCs during their active migration may be influenced by their tendency to associate with somatic cells for nutritional purposes.

Reptilia

The reptiles are an interesting group exhibiting different types of PGC migration. Risley (1933) found the PGCs to be localised in a posterior crescent in the turtle *Sternotherus* (fig. 5.4c, p. 102). He described what he called active interstitial migration of the PGCs from this site towards the gonadal anlagen through the splanchnic mesoderm and the dorsal mesentery, in much the same way as described above for the mammals. A similar situation was found by Hubert (1965, 1971*a*, *b*, 1976) in various *Lacerta* species. Hubert (1970*b*) described the formation of pseudopodia containing microfilaments on the surface of the PGCs. In *Sphenodon* Tribe & Brambell (1932) described an anterior as well as a posterior germinal crescent (fig. 5.4b, p. 102). They postulated that the PGCs of the posterior crescent would reach the gonadal anlagen by interstitial migration through the splanchnic mesoderm and dorsal mesentery, but those of the anterior crescent would enter the vitelline veins and reach the gonadal ridges by way of the vascular system – i.e. passing through the vitelline veins, the heart, the aorta and the omphalo-mesenteric artery. The presence of an anterior germinal crescent (fig. 5.4a, p. 102) and the subsequent vascular transfer of the PGCs was also described in *Mabuya* and *Chamaeleo* by Pasteels (1953). Hubert (1969) found a strictly anterior germinal crescent in *Vipera*, but in *Anguis* the PGCs were found around the entire embryonic anlage, though markedly predominating anteriorly. He observed vascular transfer of the PGCs towards the gonadal anlagen from both the anterior and the posterior extra-embryonic regions. (See the reviews by Pasteels, 1964, and Hubert, 1976.)

Summarising, it may be said that in the Chelonia the migration of the PGCs from the posterior germinal crescent towards the gonadal anlagen seems to be exclusively interstitial. Interstitial migration as well as vascular transfer is found in the Squamata. PGCs from a posterior site generally take the interstitial migration route, whereas those from an anterior site cover the much longer distance to the gonadal ridges by means of vascular transfer. Since no experimental data are available we can only guess at the mechanisms involved in the two types of migration.

General conclusions

Surveying the displacements of the PGCs in the chordates, it is evident that in the lower chordates, where the gonadal anlagen arise in the vicinity of the PGCs, no transport mechanism is required. In the vertebrates, where the PGCs have an extra-gonadal site of origin, they are both passively and actively displaced towards the gonadal anlagen. The early displacements occurring during embryogenesis are mainly passive and result from morphogenetic movements of the surrounding tissues. Although morphogenetic processes may still play a role in germ cell displacement during later phases of development, and even during gonadogenesis, the considerable distance between the location of the germ cells in the early embryo and the gonadal anlagen is bridged in the main by active migration, in which chemotactic processes play a leading role. The chemotactic attraction originates first from the dorsal mesodermal organs, leading to an accumulation of the PGCs near the dorsal root of the mesentery, and later from the epithelium of the genital ridges. The former phase has been extensively studied in the amphibians, the latter in the birds. Depending upon the absolute distance to be covered, the PGCs may make use of the blood circulation to reach the neighbourhood of the genital ridges. Vascular transfer occurs mainly in forms with meroblastic eggs, but is also related to the time of onset of blood circulation – early in the Amniota and late in the Anamnia. The use of the vascular route seems to be correlated with the nature of the passive displacements of the PGCs during early embryogenesis, i.e. whether they lead to an anterior or posterior localisation of the PGCs. Posterior localisation is usually associated with interstitial migration, whereas vascular transfer occurs in the case of anterior localisation. A careful analysis of the displacements of the PGCs during early embryogenesis seems of crucial importance. The reptiles, which show examples of both types of migration, are a very interesting group in this respect and should be studied much more extensively.

8

Phylogenetic significance of the embryological data presented

When comparing animal forms a distinction must be made between *taxonomic* relationships, which are essentially based upon similarity or dissimilarity in anatomical structure, and *phylogenetic* relationships, which are the expression of nearness or remoteness with respect to a hypothetical common ancestry. The establishment of taxonomic relationships may be hampered by the phenomenon of adaptation to a particular environment; this may lead to the appearance of similar features in the forms in question, which are however only of a secondary nature and may camouflage the true anatomical relationships. There are many examples of such adaptations, e.g. those for the aquatic, terrestrial or aerial habitat. It is in fact the fossil record which may help in ascertaining the primary or secondary nature of certain anatomical features. The comparison of the anatomical structure of widely divergent forms is rendered difficult and subjective by the arbitrariness inherent in choosing which features to emphasise.

The direct establishment of phylogenetic relationships is only possible in those exceptional cases where the common ancestry of forms can be directly demonstrated in the fossil record. In the great majority of cases phylogenetic relationships between certain forms can only be deduced indirectly, i.e. from a comparison of the anatomical structure of rare and often fragmentary fossil remains with that of present-day forms. It is therefore evident that the sharp distinction between taxonomic and phylogenetic relationships is unrealistic. The difference between them is mainly a question of what is being emphasised, the anatomical or the palaeontological evidence.

It is evident that the establishment of relatedness between two animal forms is more justified the closer the forms resemble each other. Such relatedness holds, for instance, for the comparison of two species belonging to the same genus or for two genera belonging to the same family. It becomes more and more hazardous, however, when the comparison is made between different families of the same order, different orders, subclasses or even classes. The higher the taxonomic level the less evident does the anatomical relationship become and the more dubious will be the phylogenetic relationships deduced from the very fragmentary fossil record.

How far can embryological data contribute to our insight into the taxonomic and phylogenetic relationships between different forms? We believe that embryology can make two main contributions. The first one is a *direct* contribution to a better understanding of the genesis of the ultimate anatomical structure. An accurate description of the formation of the ultimate structure during embryonic development can help immensely in understanding the complexity of that structure and in appreciating particular features of it. The causal analysis of developmental processes can moreover give us an insight into the actual mechanisms involved. It enables us to evaluate the importance of certain differences in the development and ultimate structure of the forms in question. The second contribution which embryology can make is an *indirect* one, i.e. a contribution to our insight into the phylogenetic relationships between different forms. It is a well-known fact that the more closely two forms are related, the more similar is their embryonic development. This holds particularly for early development, which is practically identical in closely related forms. That is why we can describe the early development of a certain group, e.g. the anuran or the urodele amphibians, in common terms and make a common organ anlage map valid for the whole group at a given stage of development. In general it can be said that the more closely two forms are related, the later in development do the various species-specific features become manifest. Conversely, the more remotely two forms are related, the more strongly and the more precociously will their embryonic developments diverge (cf. von Baer, 1828).

Assuming that these statements are valid, their opposite formulation will in general also hold: i.e. the earlier in development 'specific' differences representing real alterations appear between the forms being compared, the more fundamental are they and the more remotely will the forms be related. Essential differences in the early development of different forms must therefore be due to long separate evolutionary histories (cf. De Beer's (1958) concept of 'paedomorphosis').

Having arrived at this point a few words ought to be said about the validity of Haeckel's so-called biogenetic law. This states that the embryonic development of any given form represents a highly condensed recapitulation of its phylogenetic history and will therefore show transiently features that were peculiar to its ancestors. It is a well-known fact that during development animals may show features which to some extent resemble those of putative ancestors – e.g. the transient formation of gill slits in the higher vertebrates, which resembles the situation in the lower vertebrates where gill slits are actually functional, either temporarily (as in the amphibians) or permanently (as in the fishes). It must however be emphasised that this is only a superficial resemblance. De Beer (1958) sharply criticises Haeckel's reasoning and emphasises that 'phylogeny is the result of ontogeny instead of being its cause'. It should moreover be realised that the embryo must be adequately adapted to its special environment in any phase of its development in order to survive. It is now widely accepted that Haeckel's biogenetic law has no real validity.

Palaeontological evidence for the phylogeny of the vertebrates

Before discussing the embryological evidence for particular phylogenetic relationships among the recent chordates (the embryological data are naturally restricted to the living forms) we will give a brief survey of current ideas about the phylogenetic relationships among the extinct and living chordates. We have largely followed the outline given in the recent comprehensive monograph on vertebrate evolutionary history by Stahl (1974), in which the phylogenetic interpretation of the palaeontological data available today is presented as objectively as possible on the basis of the current classification of the vertebrates, and which moreover clearly indicates the basic adaptive requirements for each major group.

Although some of the forms living today show very primitive features it can be very misleading to compare them with each other. The present-day forms are the outcome of long separate evolutionary histories. They represent, as it were, the top branches of the evolutionary tree and are only connected with each other through the main branches. Only palaeontological data can give us some insight into the phylogenetic relationships, since the fossils concerned represent sections through some of the main branches. The palaeontological data available at present will therefore be briefly discussed. Since the primitive chordates consist only of soft tissues which cannot fossilise, we must restrict ourselves to the vertebrates.

The fishes

No palaeontological evidence exists on the origin of the first jawless fishes, the Ostracoderms. One can therefore only speculate about their origin by comparing them with the present-day primitive chordates, the Cephalochordata, represented by the boneless lancelot *Branchiostoma* or amphioxus and the Urochordata or Tunicata. The latter pass through a larval phase having many features in common with the early stages of development of amphioxus. It must however be realised that there is very probably a wide gap and a long evolutionary history between the tunicates and the cephalochordates, as well as between either of these two groups and the first vertebrates. The tunicates and cephalochordates are therefore placed in different subphyla of the phylum Chordata, demonstrating that their relationship is only remote. Although the anatomical differences between the urochordates and cephalochordates and the first fishes are very pronounced, among other things in the development of the central nervous system, all three groups live in an aquatic habitat. As far as environmental conditions are concerned, only adaptations to different salinities may have been required in these possible evolutionary transitions.

The oldest true vertebrates were a group of jawless fishes, the Ostracoderms. They had a shell-like, armoured skin and possessed a typical vertebrate brain and sense organs. The dermal skeleton seems to have been a later addition, since the first, middle-Ordovician remains were mostly in the form of small dermal denticles. Most of the evidence on the early Ostracoderms was found in the late-Silurian to middle-Devonian red sandstone formations in the northern hemisphere. These fossils already represent three (or four) different groups, viz. the Osteostraci, the Anaspids and the Heterostraci (the Coelolepida possibly represent larval forms of Heterostraci).

It is extremely interesting to realise that the present jawless or agnathan fishes, the Petromyzontidae and the Mixinidae, show many similarities with the most ancient Osteostraci. Although the modern jawless fishes have a much-reduced skeleton they in fact represent 'living fossils' of the most ancient vertebrates (Stensiö, 1968).

The first jawed fishes appeared by the end of the Devonian period. Although no fossil transitions between the Ostracoderms and the first jawed fishes have been found, the Osteostraci and Anaspids seem to be too specialised to be considered as possible ancestors of the jawed fishes. Only the Heterostraci could be regarded as plausible ancestors. The first jawed fishes with paired frontal appendages were the Silurian Acanthodians or spiny sharks and the Devonian, heavily armoured Placoderms. The latter form a rather heterogeneous group, however, which may have arisen polyphyletically. There are no transitional forms between the Acanthodians and Placoderms and the Devonian osteichthyan and chondrichthyan fishes. Some palaeontologists associate the Acanthodians and Placoderms with the early sharks or Chondrichthyes, whereas others defend a common ancestry of the Acanthodians and the Actinopterygii or bony fishes. This diversity of opinions demonstrates that the transition between the first jawed fishes and the main groups of modern fishes is in fact unknown (see e.g. Schaeffer, 1968).

The great development of the fishes occurred in the Devonian. In the fresh-water red sandstone formations of the mid-Devonian in Scotland representatives of all ancient and recent fishes except the marine sharks have been found. Among the Actinopterygians there was a rapid evolution from the early jawed Palaeoniscoids to the Holosteans and the Teleosteans. The Chondrichthyes or cartilaginous fishes, represented by the present-day elasmobranchs or sharks, the batoids or rays, and the holochephalians, showed an intermediate rate of evolution. The lobe-finned Rhipidistians, Coelacanths and Dipnoians or lung fishes remained very conservative and hardly evolved further up to the present time. Most authors consider the latter groups as having a common ancestry and place them together as Crossopterygians. The Coelacanths survived through the Carboniferous and Permian and showed their greatest radiation in the Triassic. Their fossil record ends abruptly in the Cretaceous. The more remarkable, therefore, was the recent discovery of the South African fish *Latimeria*, a 'living fossil' belonging to the Coelacanths. The Rhipidistians became extinct at the end of the Permian, while the Dipnoians have survived up to the present day. (See also Bertmar, 1968; Jarvik, 1968*a*.)

The origin of the Palaeoniscoids is again obscure. They declined towards the close of the Palaeozoicum and were succeeded by the more progressive Holosteans. The latter dominated in the Mesozoicum and gave rise to the teleosts. The teleosts appeared first in the Triassic as a monophyletic or polyphyletic offshoot of the Holosteans. Some authors even assume a pre-holostean origin.

The only modern survivals of the Holosteans are *Lepisosteus* and *Amia*. The present-day cartilaginous Chondrostei, the sturgeons, such as *Acipenser*, the spoonbill *Polyodon*, and the African bichir *Polypterus* also seem to have arisen from the Holosteans.

The teleosts, which were represented by only a few Triassic and Jurassic forms, diversified enormously in the Cretaceous, by which time more than a dozen different groups can already be distinguished. They ramified further in the Cenozoic era and represent the great majority of the present-day fishes.

The Actinopterygians, which replaced the Placoderms and Acanthodians and caused the decline of the Crossopterygians, did not affect the Chondrichthyes or cartilaginous fishes, which arose alongside the Palaeoniscoids in the Devonian. The origin of the Chondrichthyes is unknown. Their ancestry is sought in the bony Placoderms. Their first fossil records date from the middle Devonian, while the Palaeoniscoids were already present in the lower Devonian. The Chondrichthyes show a diversification in the Carboniferous into different lines, which may have arisen polyphyletically. The earliest Chondrichthyes are the Cladodonts of the upper Devonian, from which the Xenocanths probably evolved. Then there is the heterogeneous and aberrant group of the Bradyodonts (with permanent teeth), a separate group, the Edestids (having whirl teeth), and finally the Hybodonts, which arose early from Cladodont stock and from which the modern sharks are descended. The majority of these groups became extinct at the end of the Palaeozoicum. The Hybodonts diversified in the Mesozoicum. The modern elasmobranchs appeared in the Cretaceous. Some survivors of the Hybodonts are found in *Heterodontus* and *Chlamydoselachus*, both with a terminal mouth, while *Heptranchus*, which has seven gill slits, represents a primitive elasmobranch.

The modern elasmobranchs comprise the batoids, or rays and skates, and the aqualoids and geleoids as separate lines of sharks. Here again the question of monophyletic or polyphyletic origin is unsolved. Finally, the holocephalians, such as *Chimaera*, can be traced back to Jurassic ancestors, possibly from the same stock as the early sharks, so that they have a long separate evolutionary history.

This part of the evolutionary process is illustrated in the left half of fig. 8.1.

The amphibians

The amphibians, which were the first tetrapod 'land' animals, developed in a freshwater habitat in the Devonian period, when the Actinopterygians or bony fishes diversified rapidly. They undoubtedly emerged from the Rhipidistian lobe-fin fishes which, unlike the Coelacanths and the Dipnoians, possessed a skeleton with strong, elongated appendages which could support the body out of the water. No intermediate forms between Rhipidistians and early amphibians have been found, however. The first amphibians, the labyrinthodont Ichthyostegids found in the late Devonian deposits of Greenland, already showed all the typical tetrapod characteristics. Though able to move on land, they were still mainly aquatic and had a broad, laterally compressed tail for swimming.

One should bear in mind the many adaptations which are required for terrestrial life. Apart from the development of lungs, already found in several Devonian fishes as a probable adaptation to water of low oxygen content, the terrestrial forms needed for respiration either ribs or a pump mechanism in the floor of the mouth. The elevation of the body, required for better functioning of the respiratory mechanism, necessitated a rigid connection between the pelvic girdle and the vertebral column. The pectoral girdle had to be disconnected from the skull, which in its turn allowed the formation of an articulation between skull and vertebral column. The axial column had to be strengthened by the formation of rigid vertebrae to support the body outside the water. To prevent rapid desiccation, a beginning of keratinisation of the skin, development of a respiratory canal, and protection of the sense organs by a nasal canal and eyelids were required, while the lateral line system had to be replaced by an inner ear. When one considers the drastic nature of all these adaptations indispensable for terrestrial life, the gap between the Rhipidistian fishes and the earliest tetrapod 'land' animals, the Ichthyostegids, is very wide indeed.

Notwithstanding this wide gap the descent of the Ichthyostegids from the Rhipidistian fishes seems almost certain, mainly on the basis of the arrangement of the dermal bones in the skull (Westoll, 1943; Jarvik, 1955; Romer, 1958; Schmalhausen, 1968). This does not mean, however, that the various authors agree as to whether the tetrapods have a monophyletic or polyphyletic origin. Although Williams (1959) and Szarski (1962) advocate a unitary evolutionary history of the tetrapod traits, Holmgren (1933) and Jarvik (1955) defend a diphyletic or even polyphyletic origin (see also Thomson, 1968).

The labyrinthodont Ichthyostegids diversified in the Carboniferous into

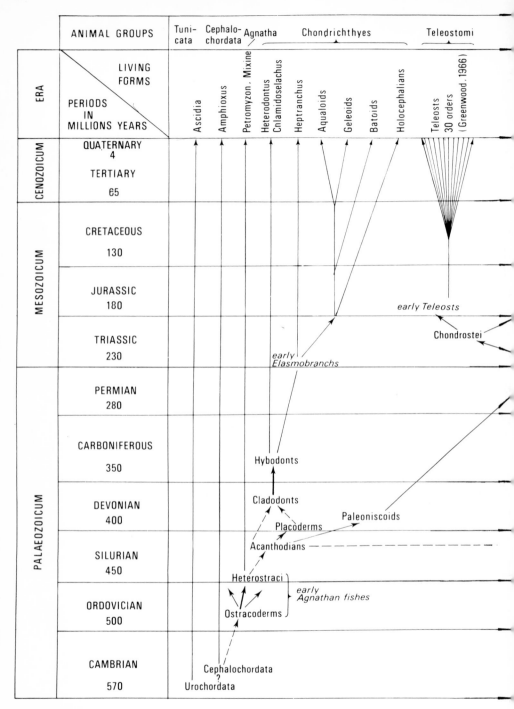

Fig. 8.1. Schematic representation of the presumed evolution of the Chordata during the Palaeozoicum, Mesozoicum and Cenozoicum on the basis of palaentological data and with reference to the present living forms.

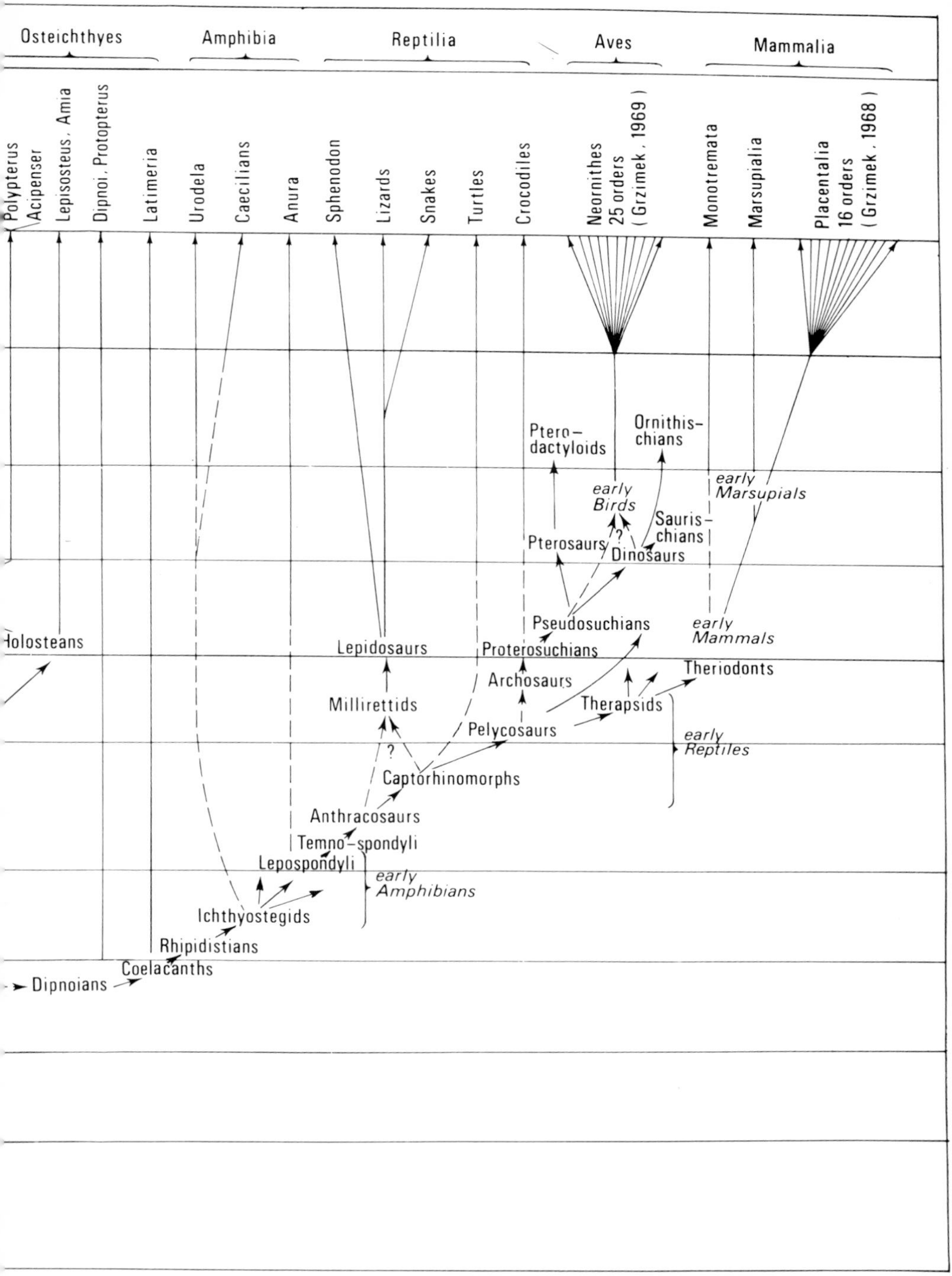

| Osteichthyes | Amphibia | Reptilia | Aves | Mammalia |

Polypterus
Acipenser
Lepisosteus. Amia
Dipnoi. Protopterus
Latimeria
Urodela
Caecilians
Anura
Sphenodon
Lizards
Snakes
Turtles
Crocodiles
Neornithes
25 orders
(Grzimek . 1969)
Monotremata
Marsupialia
Placentalia
16 orders
(Grzimek . 1968)

Holosteans

Ptero-
dactyloids

Ornithis-
chians

early
Birds

Pterosaurs

early
Marsupials

Sauris-
chians

Dinosaurs

Lepidosaurs

Pseudosuchians

early
Mammals

Proterosuchians

Archosaurs

Theriodonts

Millirettids

Therapsids

Pelycosaurs

early
Reptiles

?

Captorhinomorphs

Anthracosaurs

Temno-spondyli

Lepospondyli

early
Amphibians

Ichthyostegids

Rhipidistians

Coelacanths

Dipnoians

three groups: the specialised Aistopods, with a serpentine body often without legs, the differently specialised Nectridea, and the less specialised Lepospondyli. The Lepospondyli disappeared in the Permian. The evolution of the Labyrinthodonts was worked out by Watson (1940) and modified on the basis of more recent fossils by Romer (1947). Romer divides the Carboniferous Labyrinthodonts into the earlier Temnospondyli and the later Anthracosaurs descended from them. The latter eventually gave rise to the early reptiles in the late Carboniferous.

The possible relationships of the modern amphibians with the ancient forms are very uncertain due to lack of fossil records. Only one Triassic frog, *Triadobatrachus*, is known, the other anuran fossils being of Jurassic and Cretaceous age. Except for a single late-Jurassic urodele fossil, all representatives of this group are of the Cretaceous period. The only thing that can be said is that the present-day Coecilia or Apoda which show many primitive features, seem to be more closely related to the Urodela than to the Anura (Parker, 1956; Parsons & Williams, 1963). It must be emphasised that the modern Amphibia constitute only a remnant when compared with their representatives at the height of their development in the Carboniferous period (See fig. 8.1, pp. 134–5.).

The reptiles

The transition from amphibians to reptiles again required a number of important adaptations, of which the keratinisation or cornification of the skin and the protection of the embryo by extra-embryonic membranes and a shell are probably the most significant. We may also mention the increase in the relative size of the cerebrum and the partial partitioning of the heart that allows separation of the oxygenated blood for, amongst other organs, the brain and the deoxygenated blood for the lungs.

The forerunners of the reptiles come from early rather than late amphibian stock, the first reptiles being found when the labyrinthodont amphibians were entering their most progressive phase. The first reptiles are considered to be the mid-Carboniferous Captorhinomorphs. They diversified rapidly and gave rise to the Pelycosaurs. Their principal descendants, the Therapsids, diversified further and spread over the globe by the end of the Permian period.

The transition from amphibians to proto-reptilians is again an unsolved problem. Szarski (1968), Romer (1966) and others defend a monophyletic origin mainly on the basis of the unique development of the so-called 'land egg', but others, such as Olson (1966), suggest that the reptilian form developed from several amphibian stocks, assuming a polyphyletic origin mainly on the basis of skeletal diversity. (See also Carroll, 1969.)

On the basis of the number of temporal fenestrae in the skull the reptiles are divided into the Synapsida (single fenestra) and the Diapsida (two fenestrae). The turtles have no fenestrae and so are classed as Anapsida. Although the fossil record of the turtles starts only in the Triassic, some authors have related them directly to the early anapsid captorhinomorph Cotylosaurs, an origin which is, however, uncertain.

The synapsid Pelycosaurs very probably were the ancestors of the viviparous fish-like Ichthyosaurs, which adapted themselves fully to the aquatic habitat in the Triassic but became extinct in the Cretaceous. The main line of descent from the Pelycosaurs led to both the carnivorous and herbivorous Therapsids. They represented the first explosion of the land reptiles in the Permian period, but became extinct towards its end. They gave rise to some further branches, which, however, disappeared in the early to middle Triassic, except for the synapsid ancestors of the Mammalia, which must have arisen in an early Triassic period or even earlier. Recent investigations have made it very likely that the Therapsids acquired endo- or homeothermy (constant high body temperature) (see Desmond, 1975).

The diapsid reptiles evolved in the upper Permian and Triassic, when the Therapsids dwindled. Watson (1954) traced the diapsid reptiles back to the Anthracosaurs through the Millirettids, but Romer (1966, 1971) proposed a diphyletic origin of the diapsids from the Captorhinomorphs, one line leading via the Archosaurs to the Pseudosuchians and another via the Millirettids to the Lepidosaurs. From the Archosaurs emerged the ruling reptiles of the Mesozoic era. Along the quadrupedal line the Crocodiles emerged, primitive forms of which were found in the middle Triassic, and also the crocodile-like Phytosaurs, which were already extinct by the end of the Triassic period. Along the bipedal line emerged both the Pseudosuchians and the Jurassic Pterosaurs and Pterodactyloids, which had flight membranes. The latter were replaced by the Aves or true birds, which branched off from the Pseudo-suchians. In the reptilian ancestors of the birds endothermy was again acquired (see Desmond, 1975).

The Archosaurs gave rise, probably through the Proterosuchians, to the endothermic Dinosaurs (divided into the older Saurischians and the later Ornithischians), which became extinct at the end of the Cretaceous period. The line through the Millirettids to the Lepidosaurs led on to the modern lizards and snakes, which together form the Squamata. The snakes probably branched off from the varanoid lizards in the late Cretaceous.

In addition to the Rhynchocephalian *Sphenodon*, which again represents a 'living fossil' of the upper Permian and Triassic Lepidosaurs, the modern lizards comprise the Iguanidae, the Agamidae, the Geckonidae or geckos, the Lacertilidae or true lizards, the Draconidae or flying lizards, the Scincidae or skinks, and the legless Amphistaenids.

At the end of the Cretaceous period many reptilian groups became extinct, such as the Dinosaurs, the Pterodactylians, the Mosasaurs and several families of crocodiles, lizards and snakes. This was possibly due to unfavourable changes in climate and food supply. Only a remnant of the reptilian fauna entered the Cenozoic era.

The great diversity of reptilian groups and the ancient origin of most of them leads again to the central question as to whether they had a very early monophyletic or a slightly later polyphyletic origin. The fundamental division of the reptiles into synapsid and diapsid groups with their respective descendants, the birds and mammals, as well as the very uncertain origin of the anapsid turtles, constitute arguments for a polyphyletic origin of the reptiles. The only counter-argument concerns the possibly unique development of the so-called 'land egg'. (See fig. 8.1, pp. 134–5.)

The birds

The development of the flying birds required many adaptations. First there is the transformation of the reptilian scales into feathers; a structure transitional between scales and feathers has never been found, however. Feather formation leads to a greatly enlarged flight-supporting surface and to insulation of the body, facilitating endothermy. For the latter complete separation of the blood circulations to the lungs and the body is moreover necessary. As we have seen, these features already began to develop in the birds' reptilian forebears. Flying moreover required a reduction in weight of the skeleton and the formation of air sacs – features which were already achieved in the Pterosaurs. A further reduction in weight was attained by the loss of the teeth, while for better manoeuvrability the main weight of the animal, including the entire musculature, was shifted from the periphery to the centre of the body by the formation of long tendons. Finally, flying required a marked enlargement of the cerebellum and cerebrum.

The fossil record of the birds is very scanty. *Archaeopteryx* from the upper Jurassic limestones is surely a bird, but also has unmistakably reptilian traits which strongly resemble those of the Pseudosuchians (Galton, 1970). There is little doubt about the monophyletic origin of the birds from the endothermic Pseudosuchians. The present flightless Ratites branched off in Eocene time and are not primitive birds. In the Eocene at least half of the modern orders were already represented and the rest emerged in the Oligocene. Their maximal expansion was in the Pleistocene and a regression set in thereafter. (See fig. 8.1, pp. 134–5.)

The mammals

Mammalian development has led to a thermoregulatory mechanism different from that in birds; this was likewise already acquired by the reptilian ancestors. Mammalian status further required the formation of a secondary palate and the development of diversified tooth structures. It also called for an improvement of lung function and, as in the birds, for the full separation of lung and body blood circulations. It also included the covering of the body with fur and the formation of sweat glands for thermoregulation. Mammalian status is moreover characterised by the development of the neocortex of the cerebral hemispheres, the development of viviparity, the formation of mammary glands from modified sweat glands, and last but not least the development of placental nutrition. The last characteristic is still rudimentary in the monotremes and restricted to early development in the marsupials.

The mammals evolved from Permian therapsid reptiles which had already developed several mammalian features, including endothermy (Desmond, 1975). The early mammals were small animals of which only the teeth remain as fossil records. On the basis of tooth structure it was concluded that early therian mammals were replaced by archaic Mesozoic forms, which led to the emergence of the Mesozoic Marsupialia and the Cenozoic Placentalia. The marsupials were the most common mammals in the Mesozoic era but were for the greater part replaced by placental mammals by the end of the Cretaceous. A strong diversification of the placental mammals occurred during the transition from Mesozoicum to Cenozoicum, followed by an enormous ramification in the Eocene. Large forms appeared at the end of the Palaeocene but became extinct at the end of the Eocene. The more ancient forms were replaced by modern forms in the late Eocene to Oligocene.

After the Gondwana continent split up a regional diversification occurred on the different continents. The modern mammals originated mainly in northern Eurasia and subsequently migrated to Africa, North and South America and Australia.

There is much controversy about the question of a monophyletic or polyphyletic origin of the mammals, particularly as regards the monotremes and the marsupials (Simpson, 1959; Reed, 1960; Clemens, 1970). The Monotremata, which lay eggs and have only restricted thermoregulatory capacity, are much closer to the reptilian level than to that of the most primitive Marsupialia. The origin of the Monotremata is an open question since there are no fossil records. The Marsupialia and Placentalia, however, probably arose from a common stock, but here also there is a wide gap in the fossil record (Clemens, 1967). (See fig. 8.1, pp. 134–5.)

General considerations

One of the most striking points brought out by this palaeontological survey is the absence of fossil records at nearly all the transitional stages in the evolution of the chordates. This is in itself understandable, since successful fossilisation is a very rare event, so that generally only large populations are represented as fossils. Although adaptation to a new environment probably occurred simultaneously in various places in a given group, and several times during any one period, the population of adapted animals will always have been small, thus reducing the chances of successful fossilisation to practically zero. Moreover, transitional forms between the main groups of vertebrates have nearly always been relatively small and delicate animals which left few fossil remains – for example teeth are sole remains of the early mammals.

The other striking thing is how frequently the question of the monophyletic or polyphyletic origin of the various groups is raised. This may however, be due to the manner of approaching the problem. We have seen that, for instance, the transition from the aquatic fish to the terrestrial amphibian, or that from the reptilian to the mammalian status, required a large number of adaptations, which were often of a drastic nature. It is obvious that all these adaptations cannot have come about simultaneously, and that the transition from one status to the other must have involved a large number of successive steps and must have been a gradual process requiring much time, in the order of millions of years. Therefore, it is clear that the classification of a transitional form in one main group or another is an academic question and that the outcome must be arbitrary. The question of monophyletic or polyphyletic origin essentially concerns the absence or presence of ramifications during the transitional period. When ramifications occurred after the major transition was completed, or at least not long before the end of the transitional period, one will speak of monophyletic origin. When, on the other hand, ramifications occurred at the beginning of the transition the fossil record will suggest polyphyletic origin.

Huxley (1958), Simpson (1959) and Reed (1960) have emphasised that one can look at phylogeny in two different ways, i.e. either by asking when a given status, e.g. the amphibian or reptilian status, is actually reached, or by looking for the very first step in the realisation of such a major transition. Huxley called the first approach the 'horizontal' and the second the 'vertical' approach to phylogeny. He qualified the horizontal approach as the distinction of 'grades', the vertical one as the distinction of 'clades'. The classification of a given group with either the amphibians or the reptiles is essentially a matter of definition. Depending upon the characterisation of a major group its definition may be applicable either to a single or to more than one transitional group, suggesting a monophyletic or polyphyletic origin, respectively. On the other hand, when looking for the very first step in a major transition one is nearly always able to trace the origin of the various intermediate groups further back in history, to end up at a common origin suggesting monophyletic descent.

Although the vertical approach to phylogeny seems to open new perspectives for the understanding of the principal lines in the evolutionary process, it also has some serious drawbacks. One can always push the first step in a major transition further back in time by declaring that there has been a still earlier event, and thus a common origin of various transitional groups can always be found. This is because it is very difficult to decide what in fact is the very first step in the highly complex process of evolutionary transition. Pushing the origin of a transition further and further back in time may lead to misleading conclusions. For instance, the appearance of endothermy, which occurred in the therapsid as well as the thecodont reptiles, as an important step towards mammalian and bird status respectively, may initially have been due to a minor change in physiology or heart structure in the amphibian or fish ancestors, but it obviously does not make sense to declare that the birds and mammals are of common amphibian or fish ancestry. Agreement must therefore first be reached on the 'remoteness' of the primary step ultimately leading to the evolution of a given major group (that is, how far back in time the initial step ought to be placed). In other words, a *horizontal* criterion must be applied to the *vertical* approach to phylogeny in order to avoid meaningless or erroneous conclusions being drawn.

It should moreover be realised that the palaeontological evidence is typically stratified horizontally in a literal sense, so that it is understandable that palaeontologists favour a horizontal approach to phylogeny. However, this does not necessarily mean that a vertical approach may not be more elucidating.

Embryologists are concerned with the realisation of a particular form or function in four dimensions: that is, including the time axis. They always try to make a clear distinction between 'determination', which is the initial, usually invisible decisive event, and 'differentiation', which is the subsequent gradual realisation of a particular developmental pathway. Development can be approached horizontally by looking at the appearance of particular differentiations at given points in time, but it can also be approached vertically by seeking the initial step in the temporal chain of events leading to differentiation. It will be obvious that the experimental embryologist is inclined to favour the vertical approach as giving more insight into the actual processes involved. This will become evident in the next section.

The relevance for the phylogeny of the chordates of the embryological evidence presented

Why have we included a chapter on the phylogeny of the chordates in a monograph on germ cell development? At first sight the answer does not seem to be obvious. When studying the site and mode of origin of the PGCs in the urodele as compared with the anuran amphibians, Nieuwkoop (1947, 1950) and Sutasurya & Nieuwkoop (1974) found that the PGCs in the two groups not only have a different site of origin, but also a quite different mode of origin during early, if not very early, development. These facts, as well as other rather pronounced differences in initial development, must have their basis in very long, separate phylogenetic histories. Nieuwkoop & Sutasurya (1976) therefore questioned the proposed monophyletic origin of the Amphibia from the hypothetical Lisamphibia (Parsons & Williams, 1963) and proposed a polyphyletic origin of the Amphibia from different groups of fishes.

It therefore seems of interest to consider the evolutionary history of the entire phylum of the Chordata in the light of the available embryological evidence (see also Pasteels, 1940; De Beer, 1958). We realise, however, that the available evidence on early development (particularly its causal aspects) as well as on the origin of the PGCs is largely restricted to the amphibians and birds, while little is known about such groups as the fishes, reptiles and mammals. Nevertheless, the evidence that is available may provide some indications as to the closeness or remoteness of the phylogenetic relationships between the various groups of chordates.

So little is known about the causal factors involved in early ascidian development that a comparison of the tunicates with the cephalochordates must be restricted to pointing out the distinct similarity of their organ anlage maps. Except for a few unanswered, though rather fundamental, questions (see chapter 2, p. 38), the early development of the cephalochordates and the vertebrates seems to be based on the same general principles, so that there cannot be much doubt concerning a direct phylogenetic relationship between these two subphyla. Unfortunately no choice can be made between a direct descent of the vertebrates from the cephalochordates or a common ancestry of the two groups. At any rate the pronounced difference in level of organisation between the cephalochordates and the vertebrates, as evident e.g. in the central nervous system, suggests a phylogenetic relationship that is no more than remote.

Surveying the development of the main groups of the present-day fishes, the Agnatha and the Osteichthyes have a common, more 'primitive' type of development, whereas the development of the Chondrichthyes and the Teleostomi is more 'specialised' in character. The Teleostomi is indeed rather aberrant in that the process of gastrulation is replaced by a three-dimensional segregation of the various organ anlagen during development. Although no experimental data are available on the early development of the Agnatha or on the mode of origin of their PGCs, both processes closely resemble those in the urodele amphibians. How far this also holds for the development of the Osteichthyes, which constitute a heterogeneous group, is an open question. The available embryological evidence at least suggests a direct but remote phylogenetic relationship between the ancestors of the Agnatha and those of the Amphibia, with a possible intermediate position for the ancestors of some of the Osteichthyes. The Chondrichthyes, and particularly the Teleostomi, are very probably side branches (see, for example, Pasteels, 1940; Devillers, 1956, 1961).

As already mentioned, the pronounced differences in early development and particularly in the site and mode of origin of the PGCs between the anuran

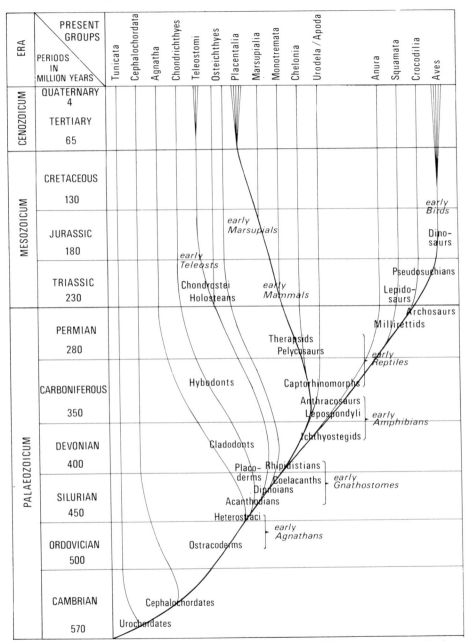

Fig. 8.2. Schematic representation of a hypothetical early bifurcation in the evolution of the Vertebrata into separate anuran–avian and urodele–mammalian lines.

and urodele amphibians plead in favour of a diphyletic origin for the amphibians, possibly from different ancestors among the Osteichthyes, so that there may be two evolutionary lines, rather than one, from the lower chordates through the fishes to the amphibians. It is not clear where the bifurcation in the evolution of the fishes may have occurred.

When we now consider the embryological evidence for phylogenetic relationships among the higher vertebrates, there are indications that the birds and the mammals are relatively far apart and represent separate evolutionary lines (see also Pasteels, 1940). This conclusion is based, among other things, on the very divergent egg types of the two groups and on the pronounced differences in their early development. When comparing the different origins of the PGCs in birds and mammals – endodermal in the former and possibly mesodermal in the latter – with that of the lower vertebrates, particularly the two groups of amphibians, avian development shows a more direct relationship with that in the anurans and mammalian development with that in the urodeles. Although this is highly speculative, it may perhaps be suggested that the divergent lines found in the amphibians continue through the reptiles to the birds and mammals respectively, the birds belonging to the anuran and the mammals to the urodele line. This suggestion is further elaborated in fig. 8.2. It places the reptiles in a crucial position, and it is therefore very unfortunate that so little is known of the early embryonic development and the origin of the PGCs in the various reptilian groups. Among the modern reptiles the crocodiles, though representing a separate group, are certainly more closely related to the Squamata than to the Chelonia. The scarce embryological evidence points towards separate lines of evolution for the two groups. Carrying speculation still further, the Chelonia may perhaps belong to the urodele–mammalian line and the Squamata to the anuran–avian line, or vice versa (see fig. 8.2).

Is there any palaeontological evidence for such an hypothesis? Jarvik (1968*b*) strongly advocates a very early diphyletic origin of the tetrapods. He places the anurans and the urodeles in the two separate main evolutionary lines in question, but suggests that all amniotes have descended from the anuran line. In our opinion, however, embryological evidence points towards a very early divergence in the ancestry of the birds and the mammals, originating already in the forebears of the ancient amphibians.

9

Concluding remarks and some perspectives for further analysis

Notwithstanding the many gaps in our knowledge the evidence presented in chapter 2 strongly suggests that epigenetic development of the mesoderm is a universal process. This conclusion seems to hold not only for all the vertebrate groups but possibly also for the primitive chordates. However, for the cephalochordates it still requires some crucial tests, while for the urochordates it is no more than a suggestion.

As we have seen in chapter 2, in the urodele amphibians the formation of the PGCs is part of the epigenetic development of the mesoderm. Since mesoderm formation in the anurans is guided by the same principles as in the urodeles and in fact comes about in almost the same manner – except for the internal position of the marginal zone in the anurans – the question arises as to why in the urodeles the PGCs develop from the presumptive lateral plate mesoderm whereas as far as we know no PGCs can develop from this or any other mesoderm in the anurans. This is undoubtedly one of the most intriguing questions as regards both mesoderm formation and PGC development.

Another important but still completely open question is the actual time and mode of determination of the PGCs in the anurans and the urodeles. A better understanding of the process of determination may contribute to bridging the gap that exists between the two groups of amphibians with respect to PGC formation.

When surveying the formation of the gonad anlagen a striking uniformity is encountered in all the vertebrate groups. Only the gonads of the amniotes show a higher level of organisation. In the lower chordates the most noticeable feature of gonad development is their segmental origin, which may in fact be considered a primitive characteristic. It is interesting to note that this feature is also found in the Coecilia, a primitive amphibian group, but is apparently absent in the Agnatha, which represent 'living fossils' of the most primitive fishes.

Except for the lower chordates, where the PGCs are formed *in situ* so that no migration is required, in all the vertebrate groups the PGCs migrate from an originally extra-gonadal or even extra-embryonic site towards the gonadal anlagen. Their initial displacements seem to be chiefly due to morphogenetic movements, while later active migration by amoeboid movement occurs. In the Anamnia the initial phase of active migration seems to occur under the influence of the dorsal mesodermal organs, while the later phase of migration is clearly under the attracting influence of the genital ridges. Depending upon the distance between the site of appearance of the PGCs and the gonadal anlagen, vascular transfer may play a part in the migration process. This is the case in those embryos where the PGCs are found in an anterior germinal crescent, as in birds and some reptiles. Here the first phase of active migration – from the endoderm into the vascular system – probably occurs through spontaneous activity of the PGCs alone, while the second phase – out of the vascular system and into the genital ridges – is clearly due to a chemotactic stimulus exerted by the gonadal epithelium, to which the PGCs react specifically. The mode of migration of the PGCs in birds differs greatly from that in mammals, demonstrating the existence of a deep division in the higher vertebrates. There is moreover in the birds the deviant appearance of the gonadal anlagen in the splanchnic rather than the somatic mesoderm. This deep division may be related to the profound differences in egg structure and early development between the two groups. It is interesting that the reptiles exhibit an intermediate position between the birds and the mammals, showing interstitial migration either with or without vascular transfer.

The writing of this monograph was prompted by the detection of fundamental differences in the site and mode of origin of the PGCs between the urodele and anuran amphibians, which led to a re-evaluation of the role of the germinal plasm in germ cell determination. As already mentioned in chapter 5 is it very difficult to envisage that one and the same cytoplasmic organelle can play such different roles in two groups of animals, i.e. that of germ cell 'determinant' in the anurans and that of a mere attendant feature of cytoplasmic differentiation in the urodeles. In our opinion this suggests the existence of a non-structural determinative factor in the germinal plasm of the anuran egg; this may also be present in the cytoplasm of the urodele egg, but there it would not be associated with any specific cytoplasmic structure. This non-structural factor would be responsible for germ cell determination in both groups, whereas the visible 'specific' structure of the germinal plasm would not be an essential feature but only an attendant phenomenon. This is a far-reaching suggestion, which certainly calls for much further experimental analysis. It is clear that the detection of the fundamental differences in germ cell formation between the two amphibian groups should keep us from making definite statements about the role of the germinal plasm in germ cell determination, and also from extending the interpretation of germ cell formation in one particular group (the anuran amphibians) to other groups of the vertebrates. The undesirability of such a generalisation (cf. Blackler, 1966) has become even more obvious since it was found that the

PGCs in birds are apparently not characterised by germinal plasm, although their formation has certain aspects in common with that in the anurans.

The situation in the mammals, where nuage material resembling germinal plasm has been found in the PGCs, could be analogous to that in the urodeles. It is very unfortunate that nothing is known about the time of appearance of this nuage material. It is also a great pity that so very little is known about the site and mode of origin of the PGCs in the fishes and reptiles.

It will be evident to every reader of this monograph that there are vast gaps in our knowledge of early embryonic development and of the place and mode of origin of the PGCs in the various groups of the chordates. Although we have a reasonable understanding of the early development of the avian embryo we know nothing about the actual mode of origin of the PGCs in this group. The knowledge of early mammalian development is still scanty but there are promising signs of considerable progress to be expected in the near future, since the culture of mammalian eggs and embryos *in vitro* has become practicable in recent years. We must say that there is simply no excuse for our ignorance about early fish and reptilian development, particularly the latter. There has been a completely unjustified neglect of these groups. The importance of a better knowledge of these groups becomes even more evident in the light of the possible phylogenetic implications of the fundamental differences in the origin of the PGCs between anuran and urodele amphibians, since this may have had its beginnings in an earlier dichotomy in the fishes and may have led to a later division in the reptiles. We hope that our admittedly very speculative suggestion of two separate phylogenetic lines running through the vertebrates from the fishes right up to the birds and mammals may stimulate the descriptive and experimental analysis of the early development of the neglected groups. They may turn out to constitute important links in these two phylogenetic lines.

For experimental research *Petromyzon* among the primitive Agnatha, and *Acipenser* and possibly some of the lung fishes among the Osteichthyes, seem particularly promising. Among the reptiles some of the representatives of the Chelonia and Lacertilia may be accessible to experimental analysis. Among the lower chordates *Branchiostoma* has turned out to be an excellent experimental animal, as shown by the elegant experiments carried out by Tung and co-workers. Unfortunately it is not very easily available.

We feel that we do not have to go into much detail in making suggestions for further analysis. We may leave the choice to the reader, who can easily find ideas in the description of the analyses carried out so far in the various groups. We believe that the problem of the site and mode of origin of the PGCs in the various groups of the chordates offers a wide field of very promising research for descriptive and experimental embryologists alike. Moreover, we are convinced that descriptive and experimental embryology can make an important contribution to our insight into the phylogenetic history of the chordates. The only unfortunate restriction is that embryology is inevitably limited to representatives of the various present-day groups. We very much hope that this monograph will stimulate new research on these intriguing problems. If our hope were to be fulfilled we would feel fully rewarded for writing it.

References

Abramowicz, H. (1913). Die Entwicklung der Gonadenanlage und Entstehung der Gonocyten bei *Triton taeniatus* (Schneid.). *Morphol. Jahrb.* **47**, 593–644.

Aisenstadt, T. B. (1975). Modern concepts of germ cell determinants. *Ontogenesis* **6**, 427–41.

Albert, J. (1976). Analyse expérimentale des interactions endo-mésodermiques au cours de l'organogenèse de l'appareil digestif chez *Rana dalmatina* Bon. (Amphibien Anoure). Dr. Sci. Thesis, University of Bordeaux, 189 pp.

Allen, B. M. (1906). The origin of the sex-cells of *Chrysemys*. *Anat. Anz.* **29**, 217–36.

Allen, B. M. (1911). The origin of the sex-cells of *Amia* and *Lepidosteus*. *J. Morphol.* **22**, 1–35.

Amanuma, A. (1957). Effect of extirpation of the presumptive intermediate mesoderm upon the differentiation of the primordial germ cells. *Zool. Mag., Tokyo* **66**, 310–13.

Amanuma, A. (1958). On the role of the dorso-caudal endoderm in the formation of the primordial germ cells. *J. Inst. Polytech. Osaka City Univ., Ser. D* **9**, 211–16.

Ancel, P. & Vintemberger, P. (1948). Recherches sur le déterminisme de la symétrie bilaterale dans l'oeuf des amphibiens. *Bull. Biol. Fr. Belg. Suppl.* **31**, 1–182.

Asayama, S. (1950). The developmental potencies of the intermediate mesoderm of *Triturus pyrrhogaster* when transplanted into orthotopic and heterotopic sites in the body wall of another embryo. *J. Inst. Polytech. Osaka City Univ. Ser. D* **1**, 13–32.

Asayama, S. (1961). Potency of lateral plate mesoderm relating to the formation of reproductive tissues. *Zool. Mag., Tokyo* **70**, 289–93.

Asayama, S. (1963). The so-called primordial germ cell in human embryos: its nature and significance on the origin of germ cells. *J. Biol., Osaka City Univ.* **14**, 1–14.

Asayama, S. (1965). A histochemical study of the so-called primordial germ cell in human embryos utilizing PAS reaction. *J. Biol., Osaka City Univ.* **16**, 53–8.

Asayama, S. (1967). Morphological and histochemical approaches to the problem of the origin of germ cells and the sex differentiation of gonads in vertebrates. *Gunma Symp. Endocrinol.* **4**, 81–99.

Asayama, S. & Amanuma, A. (1957). On the primordial germ cells of the secondary embryo induced by the organizer. *Zool. Mag., Tokyo* **66**, 280–3.

Aubry, R. (1953a). Analyse de certaines conditions expérimentales favorisant la stérilisation *ab ovo* de la grenouille rousse. *C.R. Soc. Biol.* **147**, 893–4.

Aubry, R. (1953b). Nouveaux essais de stérilisation totale des gonades de *Rana temporaria* par action des rayons ultraviolets sur le pôle inférieur de l'oeuf fécondé. *C.R. Acad. Sci., Paris* **236**, 1101–2.

Baer, K. E. von (1828). *Ueber Entwicklungsgeschichte der Thiere. Beobachtung und Reflexion.* Köningsberg.

Baker, T. G. (1963). A quantitative and cytological study of germ cells in human ovaries. *Proc. Roy. Soc., Ser. B* **158**, 417–33.

Baker, T. G. (1966). A quantitative and cytological study of oogenesis in the rhesus monkey. *J. Anat.* **100**, 761–76.

Baker, T. G. (1972). Oogenesis and ovarian development. In *Reproductive Biology*, ed. H. Balin & S. Glasser, pp. 398–437. Excerpta Medica, Amsterdam.

Balfour, F. M. (1877). The development of elasmobranch fishes. *J. Anat. Physiol.* **11**, 128–72.

Balfour, F. M. & Parker, W. N. (1882). On the structure and development of *Lepidosteus*. *Phil. Trans. Roy. Soc.* **2**, 359–441.

Balinsky, B. I. (1948). Korrelationen in der Entwicklung der Mund- und Kiemenregion und des Darmkanals bei Amphibien. *Wilhelm Roux' Arch. Entwicklungsmech. Organismen* **143**, 365–95.

Balinsky, B. I. (1966). Changes in the ultrastructure of amphibian eggs following fertilization. *Acta Embryol. Morphol. Exp.* **9**, 132–54.

Ballard, W. W. (1964). Morphogenetic movements in teleost embryos. *Amer. Zool.* **4**, 12.

Ballard, W. W. (1965). Formative movements in teleost embryos. *Amer. Zool.* **5**, 83.

Ballard, W. W. (1966*a*). The role of the cellular envelope in the morphogenetic movements of teleost embryos. *J. Exp. Zool.* **161**, 193–200.

Ballard, W. W. (1966*b*). Origin of the hypoblast in *Salmo*. I. Does the blastodisc edge turn inward? *J. Exp. Zool.* **161**, 201–10.

Ballard, W. W. (1966*c*). Origin of the hypoblast in *Salmo*. II. Outward movement of deep central cells. *J. Exp. Zool.* **161**, 211–20.

Ballard, W. W. (1968). History of the hypoblast in *Salmo*. *J. Exp. Zool.* **168**, 257–72.

Ballard, W. W. (1973*a*). Normal embryonic stages for Salmonid fishes, based on *Salmo gairdneri* Richardson and *Salvelinus fontinalis* (Mitchill). *J. Exp. Zool.* **184**, 7–26.

Ballard, W. W. (1973*b*). Morphogenetic movements in *Salmo gairdneri* Richardson. *J. Exp. Zool.* **184**, 27–48.

Ballard, W. W. (1973*c*). A new fate map for *Salmo gairdneri*. *J. Exp. Zool.* **184**, 49–74.

Ballard, W. W. (1973*d*). A re-examination of gastrulation in teleosts. *Rev. Roum. Biol.*, Sér. *Zool.* **18**, 119–35.

Ballard, W. W. & Dodes, L. M. (1968). The morphogenetic movements of the lower surface of the blastodisc in Salmonid embryos. *J. Exp. Zool.* **168**, 67–84.

Ballard, W. W. & Needham, R. G. (1964). Normal embryonic stages of *Polyodon spathula* (Walbaum). *J. Morphol.* **114**, 465–78.

Beams, H. W. & Kessel, R. G. (1974). The problem of germ cell determinants. *Int. Rev. Cytol.* **39**, 413–79.

Beard, J. (1900). The morphological continuity of the germ cells in *Raja batis*. *Anat. Anz.* **18**, 465–85.

Beard, J. (1902*a*). The germ cells of *Pristiurus*. *Anat. Anz.* **21**, 50–61.

Beard, J. (1902*b*). The germ cells. I. *Raja batis*. *Zool. Jahrb.* (*Anat.*) **16**, 615–702.

Beaumont, H. M. & Mandl, A. M. (1962). A quantitative and cytological study of oogonia and oocytes in the foetal and neonatal rat. *Proc. Roy. Soc. Ser. B* **155**, 557–79.

Belsare, D. K. (1966). Development of gonads in *Channa punctatus* Bloch (Osteichthyes: Channidae). *J. Morphol.* **119**, 467–76.

Bennett, D. (1956). Developmental analysis of a mutation with pleiotropic effects in the mouse. *J. Morphol.* **98**, 199–233.

Benoit, J. (1930). Destruction des gonocytes primaires dans le blastoderme du poulet par les rayons ultraviolets aux premiers stades du développement embryonnaire. *Proc. Int. Congr. Sex Res.* 162–70.

Berrill, N. J. (1975). Chordata: Tunicata. In *Reproduction of Marine Invertebrates*, ed. A. C. Giese & J. S. Pearse, vol. 2, pp. 241–82. Academic Press, New York & London.

Bertmar, J. (1968). Lungfish phylogeny. In *Current Problems of Lower Vertebrate Phylogeny*, ed. T. Ørvig, pp. 259–83. Wiley Interscience, New York.

Blackler, A. W. (1958). Contribution to the study of germ-cells in the Anura. *J. Embryol. Exp. Morphol.* **6**, 491–503.

Blackler, A. W. (1962). Transfer of primordial germ cells between two subspecies of *Xenopus laevis*. *J. Embryol. Exp. Morphol.* **10**, 641–51.

Blackler, A. W. (1965a). Germ cell transfer and sex ratio in *Xenopus laevis*. *J. Embryol. Exp. Morphol.* **13**, 51–61.

Blackler, A. W. (1965b). The continuity of the germ line in amphibians and mammals. *Ann. Biol.* **4**, 627–35.

Blackler, A. W. (1966). Embryonic sex cells in Amphibia. *Adv. Reprod. Physiol.* **1**, 9–28.

Blackler, A. W. (1970). The integrity of the reproductive cell line in the Amphibia. *Curr. Top. Develop. Biol.* **5**, 71–87.

Blackler, A. W. & Fischberg, M. (1961). Transfer of primordial germ cells in *X. laevis*. *J. Embryol. Exp. Morphol.* **9**, 634–41.

Blackler, A. W. & Gecking, C. A. (1972a). Transmission of sex cells of one species through the body of a second species in the genus *Xenopus*. I. Intraspecific matings. *Develop. Biol.* **27**, 376–84.

Blackler, A. W. & Gecking, C. A. (1972b). Transmission of sex cells of one species through the body of a second species in the genus *Xenopus*. II. Interspecific matings. *Develop. Biol.* **27**, 385–94.

Blandau, R. J., White, B. J. & Rumery, R. E. (1963). Observations on the movements of the living primordial germ cells in the mouse. *Fert. Steril.* **14**, 482–9.

Blocker, H. W. (1933). Embryonic history of the germ cells in *Passer domesticus* (L.). *Acta Zool., Stockholm* **14**, 111–52.

Borum, K. (1966). Oogenesis in the mouse. A study of the origin of the mature ova. *Exp. Cell Res.* **45**, 39–47.

Boterenbrood, E. C. & Nieuwkoop, P. D. (1973). The formation of the mesoderm in the urodelan amphibians. V. Its regional induction by the endoderm. *Wilhelm Roux' Arch.* **173**, 319–32.

Bounoure, L. (1939). *L'origine des cellules reproductrices et le problème de la lignée germinale* 271pp. Gauthier-Villars, Paris.

Bounoure, L., Aubry, R. & Huck, M. L. (1954). Nouvelles recherches expérimentales sur les origines de la lignée reproductrice chez la grenouille rousse. *J. Embryol. Exp. Morphol.* **2**, 245–63.

Bouvet, J. (1976). Enveloping layer and periderm of the trout embryo (*Salmo trutta fario* L.). *Cell Tissue Res.* **170**, 367–82.

Boveri, D. T. (1892). Über die Bildungsstätte der Geschlechsdrüsen und die Entstehung der Genitalkammer beim *Amphioxus*. *Anat. Anz.* **7**, 170–81.

Brambell, F. W. R. (1927). The development and morphology of the gonads of the mouse. I. The morphogenesis of the indifferent gonad and the ovary. *Proc. Roy. Soc., Ser. B* **101**, 391–409.

Brambell, F. W. R. (1960). Ovarian changes. In *Marshall's Physiology of Reproduction*, ed. A. S. Parkes, vol. 1, pp. 397–542. Longmans Green, London.

Brauer, A. (1897a). Ueber die Bildung der Keimblätter des Mesoderms und der Chorda bei Blindwühlen. *Sitz. ber. Ges. Beförd. gesammt. Naturwiss. Marburg* **2**, 21–8.

Brauer, A. (1897b). Beiträge zur Kenntniss der Entwicklungs-geschichte und der Anatomie der Gymnophionen. *Zool. Jahrb. (Anat.)* **10**, 389–472.

Bruel, M. T. (1973). Localisation des gonocytes primaires chez la jeune embryon de poulet hybride Rhode-Wyandotte. Etude histologique et expérimentale. *Arch. Anat. Histol. Embryol.* **56**, 51–64.

Bruel-Beaudenon, M. T. & Hubert, J. (1968). Remarques à propos de la lignée germinale chez certains Sauropsidés. *C.R. Soc. Biol.* **162**, 419–22.

Buehr, M. & Blackler, A. W. (1970). Sterility and partial sterility in the South African clawed toad following the pricking of the egg. *J. Embryol. Exp. Morphol.* **23**, 375–84.

Cambar, R. (1948). Recherches expérimentales sur les facteurs de la morphogenèse du mésonephros chez les Amphibiens Anoures. *Bull. Biol. Fr. Belg.* **82**, 214–85.

Cambar, R. (1952). Existe-t-il, chez les Amphibiens Anoures comme chez les Urodèles, une action inductrice de l'uretère primaire sur la crête génitale ? *C.R. Soc. Biol.* **146**, 1106–8.

Cambar, R. (1956). Les problèmes scientifiques et philosophiques de la lignée germinale chez les animaux: aspects et conceptions modernes. *Bull. Natur.* **43**, 78–103.

Cambar, R., Delbos, M. & Gipouloux, J. D. (1970). Premières observations sur l'infrastructure des cellules germinales à la fin de leur migration dans les crêtes génitales, chez les Amphibiens Anoures. *C. R. Soc. Biol.* **164**, 1686–8.

Cambar, R. & Mesnage, J. (1963). L'agénésie expérimentale du mésonephros n'influence pas le développement de la glande génitale chez les Amphibiens Anoures. *C. R. Acad. Sci., Paris* **257**, 4021–3.

Capuron, A. (1968). Analyse expérimentale de l'organogenèse urogénitale d'embryons induits par le greffe de la lèvre dorsale du blastopore chez l'amphibien urodèle *Pleurodeles waltlii* Michah. *Ann. Embryol. Morphogen.* **1**, 3–27.

Carroll, R. L. (1969). Problems of the origin of reptiles. *Biol. Rev.* **44**, 393–432.

Celestino da Costa, A. (1932a). Les gonocytes primaires chez les mammifères. *C. R. Ass. Anat., Nancy* **27** ième Réun., 198–212.

Celestino da Costa, A. (1932b). L'état actuel du problème de l'origine des cellules sexuelles. *Bull. Ass. Anat.* **27**, 1–20.

Celestino da Costa, A. (1937). Sur la migration des gonocytes primaires chez le cobaye. *C. R. Ass. Anat., Marseille*, **32** ième Réun., 1–7.

Chieffi, G. (1959). Sex differentiation and experimental sex reversal in elasmobranches. *Arch. Anat. Microsc. Morphol. Exp.* **48**, 21–36.

Chiquoine, A. D. (1954). The identification, origin, and migration of the primordial germ cells in the mouse embryo. *Anat. Rec.* **118**, 135–46.

Chiquoine, A. D. & Rothenberg, E. J. (1957). A note on alkaline phosphatase activity of germ cells in *Amblystoma* and chick embryos. *Anat. Rec.* **127**, 31–5.

Chrétien, F. C. (1966). Etude de l'origine de la migration et de la multiplication des cellules germinales chez l'embryon de lapin. *J. Embryol. Exp. Morphol.* **16**, 591–607.

Chrétien, F. C. (1968). Mise en évidence des gonocytes primordiaux de l'embryon de lapin par leur teneur en phosphatase alcaline. *Ann. Embryol. Morphogen.* **1**, 361–72.

Clark, J. M. & Eddy, E. M. (1975). Fine structural observations on the origin and associations of primordial germ cells of the mouse. *Develop. Biol.* **47**, 136–55.

Clavert, J. (1961). Le développement de la symétrie bilaterale chez les vertébrés. In *Symposium on the Germ Cells and Earliest Stages of Development*, pp. 418–39. IIE/A. Baselli, Milan.

Clawson, R. C. & Domm, L. V. (1963a). The glycogen content of the primordial germ cells in the white leghorn chick embryo. *Anat. Rec.* **145**, 218–19.

Clawson, R. C. & Domm, L. V. (1963b). Developmental changes in glycogen content of primordial germ cells in chick embryo. *Proc. Soc. exp. Biol. Med.* **112**, 533–7.

Clawson, R. C. & Domm, L. V. (1969). Origin and early migration of primordial germ cells in the chick: a study of the stages definitive primitive streak through 8 somites. *Amer. J. Anat.* **125**, 87–112.

Clayton, M. B. & Dixon, K. E. (1975). Effects of UV on cleavage of *Xenopus laevis*. *J. Exp. Zool.* **192**, 277–83.

Clemens, W. A. (1967). Origin and early evolution of marsupials. *Evolution* **22**, 1–18.

Clemens, W. A. (1970). Mesozoic mammalian evolution. *Ann. Rev. Ecol. System.* **1**, 357–90.

Clermont, Y. (1972). Kinetics of spermatogenesis in mammals: seminiferous epithelium cycle and spermatogonial renewal. *Physiol. Rev.* **52**, 198–236.

Clermont, Y. & Leblond, C. P. (1955). Spermatogenesis of man, monkey, ram and other mammals as shown by the 'periodic acid-Schiff' technique. *Amer. J. Anat.* **96**, 229–53.

Clermont, Y. & Leblond, C. P. (1959). Differentiation and renewal of spermatogonia in the monkey *Macacus rhesus*. *Amer. J. Anat.* **104**, 237–72.

Coggins, L. W. (1973). An ultrastructural and radio-autographic study of early oogenesis in the toad *Xenopus laevis*. *J. Cell Sci.* **12**, 71–93.

Comings, D. E. & Okada, T. A. (1972). The chromatoid body in mouse spermatogenesis: evidence that it may be formed by the extrusion of nucleolar components. *J. Ultrastruct. Res.* **39**, 15–32.

Conklin, E. G. (1905*a*). The organization and cell-lineage of the ascidian egg. *J. Acad. Natur. Sci. Philadelphia* **13**, 1–119.

Conklin, E. G. (1905*b*). Organ-forming substances in the eggs of ascidians. *Biol. Bull.* **8**, 205–30.

Conklin, E. G. (1932). The embryology of amphioxus. *J. Morphol.* **54**, 69–151.

Conklin, E. G. (1933). The development of isolated and partially separated blastomeres of amphioxus. *J. Exp. Zool.* **64**, 303–51.

Cuminge, D. & Dubois, R. (1971). Etude ultrastructurale et autoradiographique de l'organogenèse sexuelle précoce chez l'embryon de poulet. *Exp. Cell Res.* **64**, 243–58.

Czołowska, R. (1969). Observations on the origin of the 'germinal cytoplasm' in *Xenopus laevis*. *J. Embryol. Exp. Morphol.* **22**, 229–51.

Czołowska, R. (1972). The fine structure of the 'germinal cytoplasm' in the egg of *Xenopus laevis*. *Wilhelm Roux' Arch. Entwicklungsmech. Organismen* **169**, 335–44.

Dalcq, A. (1935). *L'organisation de l'oeuf chez les Chordés. Etude d'embryologie causale*, 332pp. Gauthier-Villars, Paris.

Dalcq, A. (1938). *Form and Causality in Early Development*, 191 pp. Cambridge University Press, Cambridge.

D'Ancona, U. (1950). Détermination et différenciation du sexe chez les poissons. In Colloque sur la Différenciation Sexuelle chez les Vertébrés, CNRS, Paris. *Arch. Anat. Microsc. Morphol. Exp.* **39**, 274–94.

Daniel, J. C. (1976). The first potential i.c.m. cell during cleavage of the rabbit ovum. *Wilhelm Roux' Arch. Develop. Biol.* **179**, 249–250.

Dantschakoff, V. (1932*a*). Les cellules génitales et leur continuité. *Rev. Gén. Sci.* **43**, 295–309.

Dantschakoff, V. (1932*b*). Keimzelle und Gonade. II. Ganzheit des Gewebekomplexes als Faktor in der Entwicklung der Gonade. *Z. Zellforsch. Mikrosk. Anat.* **15**, 581–644.

Dantschakoff, V. (1933). Keimzelle und Gonade. V. Sterilisierung der Gonaden im Embryo mittels Röntgenstrahlen. *Z. Zellforsch. Mikrosk. Anat.* **18**, 56–109.

Dantschakoff, V. (1936). Keimzelle und Gonade. VII. Gefässmechanismus als Faktor der Gonadenlokalisation und der primären Asymmetrie beim Hühnchen. *Z. Zellforsch. Mikrosk. Anat.* **24**, 64–85.

Dantschakoff, V. (1950). La différenciation du sexe chez les vertébrés. *Arch. Anat. Microsc. Morphol. Exp.* **39**, 367–93.

De Beer, G. (1958). *Embryos and Ancestors*, 3rd edn, 197pp. Clarendon Press, Oxford.

De Smet, W. M. A. (1970). The germ cells of *Polypterus* (Brachiopterygii, Pisces). *Acta Morphol. Neerl.-Scand.* **8**, 133–41.

Delbos, M., Gipouloux, J. D. & Cambar, R. (1971). Observations sur l'infrastructure des différents types cellulaires constituant la gonade larvaire chez la grenouille agile *Rana dalmatina* Bon. (Amphibien Anoure). *C. R. Acad. Sci., Paris* **272**, 2372–4.

Desmond, A. J. (1975). *The Hot Blooded Dinosaurs: A Revolution in Palaeontology*, 238pp. Blond & Briggs, London.

Deuchar, E. M. (1972). *Xenopus laevis* and developmental biology. *Biol. Rev.* **47**, 37–112.

Devillers, C. (1956). Les aspects caractéristiques de la prémorphogénèse dans l'oeuf des téléostéens. L'origine de l'oeuf télolécitique. *Ann. Biol.* **32**, 437–56.

Devillers, C. (1961). Structural and dynamic aspects of the development of the teleostean egg. *Adv. Morphogen.* **1**, 379–428.

Didier, E. & Fargeix, N. (1974). Etude comparée chez les oiseaux de la colonisation des gonades par les cellules germinales au cours du développement normal et après reduction expérimentale du territoire gonadique. *Arch. Anat. Histol. Embryol.* **56**, 33–50.

Didier, E. & Fargeix, N. (1976a). Aspects quantitatifs du peuplement des gonades par les cellules germinales chez l'embryon de caille (*Coturnix coturnix japonica*). *J. Embryol. Exp. Morphol.* **35**, 637–48.

Didier, E. & Fargeix, N. (1976b). Germinal population of gonads in some chimerical embryos obtained by connecting pieces of Japanese quail and domestic chick blastoderms. *Experientia* **32**, 1333–4.

Didier, E., Fargeix, N. & Bergeaud, Y. (1974). Analyse expérimentale de la régulation du nombre des cellules germinales après déficience gonadique chez le poulet. *J. Embryol. Exp. Morphol.* **32**, 619–35.

Dubois, R. (1962). Sur la stérilization de l'embryon de poulet par irradiation aux rayons X du croissant germinal extra-embryonnaire. *Arch. Anat. Microsc. Morphol. Exp.* **51**, 85–94.

Dubois, R. (1965). La lignée germinale chez les reptiles et les oiseaux. *Ann. Biol.* **4**, 637–66.

Dubois, R. (1967). Localisation et migration des cellules germinales du blastoderme non incubé de poulet d'après les résultats de culture *in vitro*. *Arch. Anat. Microsc. Morphol. Exp.* **56**, 245–64.

Dubois, R. (1968). La colonisation des ébauches gonadiques par les cellules germinales de l'embryon de poulet en culture *in vitro*. *J. Embryol. Exp. Morphol.* **20**, 189–213.

Dubois, R. (1969). Le mécanisme d'entrée des cellules germinales primordiales dans le réseau vasculaire, chez l'embryon de poulet. *J. Embryol. Exp. Morphol.* **21**, 255–70.

Dubois, R. & Croisille, Y. (1970). Germ-cell line and sexual differentiation in birds. *Phil. Trans. Roy. Soc. Ser. B* **259**, 73–89.

Dubois, R. & Cuminge, D. (1968). Sur l'aspect ultrastructural et histochimique des cellules germinales de l'embryon de poulet. *Ann. Histochim.* **13**, 33–50.

Dubois, R. & Cuminge, D. (1970). Les propriétés sécrétrices et excrétrices de l'epithelium germinatif de l'embryon de poulet: étude morphologique et dynamique par l'autoradiographie au microscope électronique. *Ann. Biol.* **9**, 479–90.

Dubois, R. & Cuminge, D. (1974). Chimiotactisme et organisation biologique. Etude de l'installation de la lignée germinale dans les ébauches gonadiques chez l'embryon de poulet. *Ann. Biol.* **13**, 241–58.

Dubois, R. & Cuminge, D. (1975). Les bases biochimiques de l'attraction chimiotactique des cellules germinales chez l'embryon de poulet. *Develop. Growth Diff.* **17**, 301–2.

Dufaure, J. P. & Hubert, J. (1965). Origine et migration des gonocytes primordiaux chez l'embryon de lézard vivipare (*Lacerta vivipara* Jacquin). *C. R. Acad. Sci., Paris* **261**, 237–40.

Dulbecco, R. (1946). Sviluppo di gonadi in assenza di cellule sessuali nell embrioni di pollo. Sterilizzazione completa mediante esposizione a raggi γ allo stadio di linea primitiva. *R. C. Accad. Naz. Lincei, Ser.* 8 **1**, 1211–13.

Dulbecco, R. (1948). Azione dei raggi γ del radio sullo sviluppo della gonade e sui caratteri somatici del sesso nell'embrione di pollo. *R. C. Accad. Naz. Lincei, Ser.* 8 **2**, 1–20.

Dziadek, M. & Dixon, K. E. (1975). Mitoses in presumptive primordial germ cells in post-blastula embryos of *Xenopus laevis*. *J. Exp. Zool.* **192**, 285–91.

Dziadek, M. & Dixon, K. E. (1977). An autoradiographic analysis of nucleic acid synthesis in the presumptive primordial germ cells of *Xenopus laevis*. *J. Embryol. Exp. Morphol.* **37**, 13–31.

Eddy, E. M. (1970). Cytochemical observations on the chromatoid body of the male germ cells. *Biol. Reprod.* **2**, 114–28.

Eddy, E. M. (1974). Fine structural observations on the form and distribution of nuage in germ cells of the rat. *Anat. Rec.* **178**, 731–58.

Eddy, E. M. (1975). Germ plasm and the differentiation of the germ cell line. *Int. Rev. Cytol.* **43**, 229–80.

Eddy, E. M. & Ito, S. (1971). Fine structural and radioautographic observations on dense perinuclear cytoplasmic material in tadpole oocytes. *J. Cell Biol.* **40**, 90–108.

Eigenmann, C. H. (1891). On the precocious segregation of the sex cells in *Micrometus aggregatus* Gibbons. *J. Morphol.* **5**, 481–92.

Essenberg, J. M. & Sreyda, A. J. (1939). The effect of the destruction of the germinal crescent on the origin of the germ cells and the development of the gonads in the domestic fowl. *West. J. Surg.* **47**, 318–27.

Everett, N. B. (1943). Observational and experimental evidences relating to the origin and differentiation of the definitive germ cells in mice. *J. Exp. Zool.* **92**, 49–91.

Everett, N. B. (1945). The present status of the germ cell problem in vertebrates. *Biol. Rev. Cambridge Phil. Soc.* **20**, 45–55.

Eyal-Giladi, H. (1954). Dynamic aspects of neural induction in Amphibia. *Arch. Biol.* **65**, 179–259.

Eyal-Giladi, H. (1969). Differentiation potencies of the young chick blastoderm as revealed by different manipulations. I. Folding experiments and position effects of the culture medium. *J. Embryol. Exp. Morphol.* **21**, 177–92.

Eyal-Giladi, H. (1970*a*). Differentiation potencies of the young chick blastoderm as revealed by different manipulations. II. Localized damage and hypoblast removal experiments. *J. Embryol. Exp. Morphol.* **23**, 739–49.

Eyal-Giladi, H. (1970*b*). Competence and induction in the pregastrular chick blastoderm (a study of Siamese twins formed in folded blastoderms). *Ann. Embryol. Morphogen.* **3**, 133–43.

Eyal-Giladi, H. & Kochav, S. (1976). From cleavage to primitive streak formation: a complementary normal table and a new look at the first stages of the development of the chick. *Develop. Biol.* **49**, 321–37.

Eyal-Giladi, H., Kochav, S. & Menashi, M. K. (1976). On the origin of the primordial germ cells in the chick embryo. *Differentiation* **6**, 13–16.

Eyal-Giladi, H. & Spratt, N. T. (1965). The embryo-forming potencies of the young chick blastoderm. *J. Embryol. Exp. Morphol.* **13**, 267–73.

Eyal-Giladi, H. & Wolk, M. (1970). The inducing capacities of the primary hypoblast as revealed by transfilter induction studies. *Wilhelm Roux' Arch. Entwicklungsmech. Organismen* **165**, 226–41.

Falconer, D. S. & Avery, P. J. (1978). Variability of chimaeras and mosaics. *J. Embryol. Exp. Morphol.* **43**, 195–219.

Falin, L. I. (1969). The development of genital glands and the origin of germ cells in human embryogenesis. *Acta Anat.* **72**, 195–232.

Fargeix, N. (1967). Stérilité partielle d'embryons de canard développés à partir de moités antérieures et postérieures de blastoderme isolées au stade non incubé. *C. R. Acad. Sci., Paris* **265**, 133–6.

Fargeix, N. (1969). Les cellules germinales du canard chez des embryons normaux et des embryons de régulation. Etude des jeunes stades du développement. *J. Embryol. Exp. Morphol.* **22**, 477–503.

Fargeix, N. (1970). La colonisation des gonades par les cellules germinales chez des embryons de régulation. Etude chez le canard. *Ann. Embryol. Morphogen.* **3**, 107–31.

Fargeix, N. (1975). La localisation du matériel germinal dans le blastoderme non incubé de l'oeuf de cane: étude à l'aide des irradiations aux rayons X. *J. Embryol. Exp. Morphol.* **34**, 171–7.

Fargeix, N. (1976). Régulation du nombre des gonocytes dans les ébauches gonadiques de l'embryon de canard après destruction partielle du stock initial des cellules germinales. *C. R. Acad. Sci., Paris* **282**, 305–8.

Farinella-Feruzza, N. (1959). Fenomeni di induzione nelle ascidie. *Boll. Zool.* **26**, 357–63.

Farinella-Feruzza, N. & Reverberi, G. (1969). Gigantic larvae of ascidians from two fused eggs. *Acta Embryol. Exp.* **2**, 281–90.

Fawcett, D. W. (1972). Observations on cell differentiation and organelle continuity in spermatogenesis. In *The Genetics of the Spermatozoon*, ed. R. A. Beatty & S. Gluecksohn-Waelsch, pp. 37–68. Beatty & Gluecksohn-Waelsch, Edinburgh & New York.

Fawcett, D. W., Eddy, E. M. & Phillips, D. M. (1970). Observations on the fine structure and relationships of the chromatoid body in mammalian spermatogenesis. *Biol. Reprod.* **2**, 129–53.

Fischiarolo, G. (1960). La formazione di cellule germinali negli Anfibi anuri dopo distribuzione de blastomeri vegetativi della blastula. *R. C. Accad. Naz. Lincei, Ser.* 8 **28**, 519–21.

Flynn, T. T. & Hill, J. P. (1947). The development of Monotremata. VI. The later stages of cleavage and the formation of the primary germ-layers. *Trans. Zool. Soc.* **26**, 1–151.

Fontaine, J. & Le Douarin, N. M. (1977). Analysis of endoderm formation in the avian blastoderm by the use of quail–chick chimaeras. The problem of the neurectodermal origin of the cells of the A.P.U.D. series. *J. Embryol. Exp. Morphol.* **41**, 209–22.

Franchi, L. L., Mandl, A. M. & Zuckerman, S. (1962). The development of the ovary and the process of oogenesis. In *The Ovary*, ed. S. Zuckerman, vol. 1, pp. 1–88. Academic Press, New York & London.

Fraser, R. C. (1954). Studies on the hypoblast of the young chick embryo. *J. Exp. Zool.* **126**, 349–99.

Fujimoto, T., Miyayama, Y. & Fuyuta, M. (1977). The origin, migration and fine morphology of human primordial germ cells. *Anat. Rec.* **188**, 315–30.

Fujimoto, T., Ninomiya, T. & Ukeshima, A. (1976*a*). Observations of the primordial germ cells in blood samples from the chick embryo. *Develop. Biol.* **49**, 278–82.

Fujimoto, T., Ukeshima, A. & Kiyofuji, R. (1975). Light- and electron-microscopic studies on the origin and migration of the primordial germ cells in the chick. *Acta Anat. Nippon.* **50**, 22–40.

Fujimoto, T., Ukeshima, A. & Kiyofuji, R. (1976*b*). The origin, migration and morphology of the primordial germ cells in the chick embryo. *Anat. Rec.* **185**, 139–54.

Fukuda, T. (1976). Ultrastructure of primordial germ cells in human embryo. *Virchows Arch., Ser. B Cell Pathol.* **20**, 85–9.

Fuyuta, M., Miyayama, Y. & Fujimoto, T. (1974). Histochemical identification of primordial germ cells in human embryos by PAS reaction. *Okajimas Folia Anat. Jap.* **51**, 251–62.

Gaillard, P. J. (1950). Sex cell formation in explants of the foetal human ovarian cortex. I and II. *Proc. Kon. Ned. Akad. Wetensch.* **53**, 1300–47.

Gaillard, P. J. (1952). Regeneratie van ovarium parenchym *in vitro*. *Genootsch. Nat. Genees- en Heelk. Amsterdam, Ser. II* **19**, 34–5.

Gallera, J. & Castro-Correia, J. (1964). Les roles respectifs de l'ectophylle et de l'entophylle dans la détermination de la symétrie bilatérale chez le poulet. *C. R. Ass. Anat.* **121**, 130–3.

Gallera, J. & Nicolet, G. (1969). Le pouvoir inducteur de l'endoblaste présumptif contenu dans la ligne primitive jeune de poulet. *J. Embryol. Exp. Morphol.* **21**, 108–18.

Gallien, L. & Durocher, M. (1957). Table chronologique du développement chez *Pleurodeles waltlii*. *Bull. Biol. Fr. Belg.* **91**, 97–114.

Galton, P. M. (1970). Ornithischian dinosaurs and the origin of birds. *Evolution* **24**, 448–62.

Gamo, H. (1961*a*). On the polyinvagination in embryo of a teleost, the medaka, *Oryzias latipes*. *Jap. Soc. Sci. Fish.* **27**, 236–7.

Gamo, H. (1961*b*). On the origin of germ cells and formation of gonad primordia in the medaka, *Oryzias latipes*. *Jap. J. Zool.* **13**, 101–15.

Gamo, H. (1961*c*). Bilateral asymmetry of gonad primordia of a fresh-water teleost, the medaka, *Oryzias latipes*. *Jap. J. Ichthyol.* **8**, 83–5.

Gardner, R. L. (1972). An investigation of inner cell mass and trophoblast tissue following their isolation from the mouse blastocyst. *J. Embryol. Exp. Morphol.* **28**, 279–312.

Gardner, R. L. (1975). Origins and properties of trophoblast. In *Immunobiology of Trophoblast*, ed. R. G. Edwards, C. W. S. Howe & M. Johnson, pp. 43–65. Cambridge University Press.

Gardner, R. L. & Papaioannou, V. E. (1975). Differentiation in the trophectoderm and inner cell mass. In *The Early Development of Mammals, British Society for Developmental Biology Symposium* 2, ed. M. Balls & A. E. Wild, pp. 107–34. Cambridge University Press.

Gardner, R. L. & Rossant, J. (1976). Determination during embryogenesis. In *Embryogenesis in Mammals*, ed. K. Elliott & M. O'Connor. *Ciba Foundation Symposium* 40, pp. 5–25. Elsevier, Amsterdam.

Ginsburg, A. S. & Detlaff, T. A. (1955). *The Embryonic Development of the Sturgeon*, pp. 1–76. USSR Academy of Sciences. (In Russian.)

Giorgi, P. P. (1974). Germ cell migration in toad (*Bufo bufo*): effect of ventral grafting of embryonic dorsal regions. *J. Embryol. Exp. Morphol.* **31**, 75–87.

Gipouloux, J. D. (1962). Mise en évidence du 'cytoplasme-germinal' dans l'oeuf et l'embryon du Discoglosse: *Discoglossus pictus* Otth. (Amphibien, Anoure). *C. R. Acad. Sci., Paris* **254**, 2433–5.

Gipouloux, J. D. (1967). Recherches expérimentales sur l'origine et la migration des cellules germinales et sur la morphogenèse de la glande génitale chez les Amphibien Anoures. Dr. Sci. Thesis, University of Bordeaux, 210 pp.

Gipouloux, J. D. (1971). Effects de l'extrusion totale ou partielle du cytoplasme germinal au cours des premiers stades de la segmentation sur la fertilité des larves d'Amphibiens Anoures. *C. R. Acad. Sci., Paris* **273**, 2627–9.

Gipouloux, J. D. (1972). Régulation du nombre des cellules germinales après un excès initial de celles-ci dans la gonade larvaire des Amphibiens Anoures. *C. R. Acad. Sci., Paris* **274**, 2229–31.

Gipouloux, J. D. (1973). Conséquences de l'augmentation expérimentale du nombre des cellules médullaires sur l'organogenèse précoce de la gonade embryonnaire du Crapaud commun, *Bufo bufo* L. (Amphibien Anoure). *C. R. Acad. Sci., Paris* **277**, 1217–19.

Gipouloux, J. D. (1975). Cytoplasme germinale et détermination germinale chez les Amphibiens Anoures. *Ann. Biol.* **14**, 475–87.

Goldsmith, J. B. (1935). The primordial germ cells of the chick. I. The effect on the gonad of complete and partial removal of the 'germinal crescent' and of the removal of other parts of the blastodisc. *J. Morphol.* **58**, 537–53.

Gomes-Ferreira, J. V. (1956). Quelques observations sur l'origine et la migration des gonocytes primaires chez l'embryon du cobaye. *C. R. Ass. Anat.* **43**, 365–73.

Gomes-Ferreira, J. V. (1957). Les gonocytes primaires chez l'embryon du cobaye. Quelques observations sur leur origine et leur migration. *C. R. Soc. Biol.* **151**, 1486–7.

Gondos, H. & Hobel, C. J. (1971). Ultrastructure of germ cell development in the human fetal testis. *Z. Zellforsch.* **119**, 1–20.

Grant, P., Wacaster, J. & Turner, P. Y. (1976). Ultraviolet irradiation delays amphibian gastrulation and prevents neural induction. *J. Cell Biol.* **70**, 363.

Greenwood, P. H., Rosen, D. E. Weitzman, S. H. & Myers, G. S. (1966). Phyletic studies of Teleostean fishes with a provisional classification of living forms. *Bull. Amer. Mus. Nat. Hist.* **131**, 393–403.

Grobstein, C. (1952). Intra-ocular growth and differentiation of clusters of mouse embryonic shields cultured with and without primitive endoderm and in the presence of possible inductors. *J. Exp. Zool.* **119**, 355–80.

Grunz, H. (1975). Studies on early embryonic induction and differentiation. In *New Approaches to the Evaluation of Abnormal Embryonic Development, Second Symposium on Prenatal Development*, ed. D. Neubert & H. J. Merker, pp. 792–803. Georg Thieme, Stuttgart.

Grunz, H., Multier-Lajous, A. M., Herbst, P. & Arkenberg, G. (1975). The differentiation of isolated amphibian ectoderm with and without treatment with an inductor. *Wilhelm Roux' Arch. Develop. Biol.* **178**, 277–84.

Grzimek, B. (1968–72). *Enzyklopädie des Tierreiches*, vols. 7–9 (birds), 10–13 (mammals). Kindler Verlag A. G.

Gurdon, J. B. (1974). The genome in specialized cells, as revealed by nuclear transplantation in Amphibia. In *The Cell Nucleus*, ed. H. Busch, vol. 1, pp. 471–89. Academic Press, New York & London.

Hama, T. (1949). Explantation of the urodelean organizer and the process of morphological differentiation attendant upon invagination. *Proc. Jap. Acad.* **25**, 4–11.

Hamburger, V. & Hamilton, H. (1951). A series of normal stages in the development of the chick embryo. *J. Morphol.* **88**, 49–92.

Hamilton, W. J. & Mossman, H. W. (1972). *Hamilton, Boyd & Mossman's Human Embryology*, 4th edn, 493pp. Heffer, Cambridge.

Hann, H. W. (1927). The history of the germ cells of *Cottus bairdii* Girard. *J. Morphol. Physiol.* **43**, 427–97.

Hara, K. (1977). The cleavage pattern of the axolotl egg studied by cinematography and cell counting. *Wilhelm Roux' Arch. Develop. Biol.* **181**, 73–87.

Hardisty, M. W. (1971). Gonadogenesis, sex differentiation and gametogenesis. In *The Biology of the Lampreys*, ed. M. W. Hardisty & J. C. Potter, vol. 1, pp. 295–359. Academic Press, New York & London.

Hardisty, M. W. & Cosh, J. (1966). Primordial germ cells and fecundity. *Nature, Lond.* **210**, 1370–1.

Harrison, R. G. (1969). Harrison stages and description of the normal development of the spotted salamander, *Ambystoma punctatum* (Linn.). In *Organization and Development of the Embryo*, ed. S. Wilens, pp. 44–66. Yale University Press, New Haven.

Hatschek, B. (1881). Studien über die Entwicklung des Amphioxus. *Arb. Zool. Inst., Wien* **4**, 1–88.

Hatschek, B. (1888). Über die Schichtenbau von Amphioxus. *Anat. Anz.* **3**, 662–7.

Hegner, R. W. (1911). Germ-cell determinants and their significance. *Amer. Natur.* **45**, 385–97.

Henderson, I. F. & Henderson, W. D. (1953). *A Dictionary of Scientific Terms*, 5th edn, 506pp. Oliver & Boyd, Edinburgh.

Hill, J. P. (1910). The early development of the Marsupialia with special reference to the native cat (*Dasyurus viverrinus*). *Quart. J. Microsc. Sci.* **56**, 1–166.

Hilscher, W. (1967). DNA synthesis, proliferation and regulation of the spermatogonia in the rat. *Arch. Anat. Microsc. Morphol. Exp.* **56**, 75–84.

Hilscher, W. & Hilscher, B. (1976). Kinetics of the male gametogenesis. *Andrologia* **8**, 105–16.

Hoessels, E. L. M. J. (1957). Evolution de la plaque préchordale d'*Ambystoma mexicanum*: sa différenciation propre et sa puissance inductrice pendant la gastrulation. PhD Thesis, University of Utrecht, The Netherlands, 71pp.

Hoffmann, C. K. (1896). Beiträge zur Entwicklungsgeschichte der Selachii. *Morphol. Jahrb.* **24**, 209–86.

Hogan, J. C. (1973). The fate and fine structure of primordial germ cells in the teleost, *Oryzias latipes*. *J. Cell Biol.* **59**, 146 (abstr.).

Hogarth, K. & Dixon, K. E. (1976). Protein synthesis and germ plasm in cleavage embryos of *Xenopus laevis*. *J. Exp. Zool.* **198**, 429–35.

Holmgren, N. (1933). On the origin of the tetrapod limb. *Acta Zool., Stockholm* **4**, 1–295.

Holtfreter, J. (1938a). Differenzierungspotenzen isolierter Teile der Urodelen-gastrula. *Wilhelm Roux' Arch. Entwicklungsmech. Organismen* **138**, 522–656.

Holtfreter, J. (1938b). Differenzierungspotenzen isolierter Teile der Anurengastrula. *Wilhelm Roux Arch. Entwicklungsmech. Organismen* **138**, 657–738.

Holtfreter, J. (1943). A study of the mechanics of gastrulation. I. *J. Exp. Zool.* **94**, 261–318.

Holtfreter, J. (1944). A study of the mechanics of gastrulation. II. *J. Exp. Zool.* **95**, 171–212.

Holtfreter-Ban, H. (1965). Differentiation capacities of Spemann's organizer investigated in explants of diminishing size. PhD Thesis, University of Rochester, Ann Arbor, Michigan, 252pp.

Houillon, C. (1956). Recherches expérimentales sur la dissociation médullo-corticale dans d'organogénèse des gonades chez le triton, *Pleurodeles waltlii* Michah. *Bull. Biol. Fr. Belg.* **90**, 359–445.

Hubert, J. (1962). Etude histologique des jeunes stades du développement embryonnaire du lézard vivipare (*Lacerta vivipara* Jacquin.). *Arch. Anat. microsc. Morphol. expér.* **51**, 11–26.

Hubert, J. (1965). Première confirmation expérimentale de l'origine extra-embryonnaire des gonocytes primordiaux chez le lézard vivipare (*Lacerta vivipara* Jacquin). *C. R. Acad. Sci., Paris* **261**, 4505–8.

Hubert, J. (1968). A propos de la lignée germinale chez 2 reptiles: *Anguis fragilis* L. et *Vipera aspis* L. *C. R. Acad. Sci., Paris* **266**, 231–3.

Hubert, J. (1969). Localisation précoce et mode de migration des gonocytes primor-diaux chez quelques reptiles. *Ann. Embryol. Morphogen.* **2**, 479–94.

Hubert, J. (1970*a*). Etude cytologique et cytochimique des cellules germinales des reptiles au cours du développement embryonnaire et après la naissance. *Z. Zellforsch. Mikrosk. Anat.* **107**, 249–64.

Hubert, J. (1970*b*). Ultrastructure des cellules germinales au cours du développement embryonnaire du lézard vivipare (*Lacerta vivipara* Jaquin). *Z. Zellforsch. Mikrosk. Anat.* **107**, 265–83.

Hubert, J. (1971*a*). La localisation extra-embryonnaire des cellules germinales chez l'embryon de lézard vivipare (*Lacerta vivipara* Jacquin). *Experientia* **27**, 1463–4.

Hubert, J. (1971*b*). Localisation extra-embryonnaire des gonocytes chez l'embryon d'orvet (*Anguis fragilis* L.). *Arch. Anat. Microsc. Morphol. Exp.* **60**, 261–8.

Hubert, J. (1976). La lignée germinale chez les reptiles au cours du développement embryonnaire. *Ann. Biol.* **15**, 547–65.

Huck, M. L. & Aubry, R. (1952). Sur les signes d'une régulation du germen dans les gonades de *Rana temporaria* après destruction étendue du déterminant germinal. *C. R. Acad. Sci., Paris* **234**, 1222–4.

Hughes, G. C. (1963). The population of germ cells in the developing female chick. *J. Embryol. Exp. Morphol.* **11**, 513–36.

Humphrey, R. R. (1925). The primordial germ cells of *Hemidactylium* and other Amphibia. *J. Morphol. Physiol.* **41**, 1–43.

Humphrey, R. R. (1927). Extirpation of the primordial germ cells in *Amblystoma*: its effect upon the development of the gonad. *J. Exp. Zool.* **49**, 363–99.

Humphrey, R. R. (1928). The developmental potencies of the intermediate mesoderm of *Amblystoma* when transplanted into ventro-lateral sides in other embryos: the primordial germ cells of such grafts and their role in the development of a gonad. *Anat. Rec.* **40**, 67–90.

Humphrey, R. R. (1929). The early position of the primordial germ cells in urodeles: evidence from experimental studies. *Anat. Rec.* **42**, 301–14.

Huxley, J. S. (1958). Evolutionary processes and taxonomy with special reference to grades. *Uppsala Univ. Arsskrift.* **6**, 21–39.

Ignatieva, G. M. (1960). Regional induction by the chordomesoderm in sturgeon. *Dokl. Akad. Nauk SSSR* **134**, 706–10.

Ignatieva, G. M. (1962). Inductive properties of the chordomesodermal rudiment before onset of invagination and regulation of defects of this rudiment in sturgeon embryos. *Dokl. Akad. Nauk SSSR* **139**, 588–91.

Ijiri, K. I. (1976). Stage-sensitivity and dose–response curve of UV effect on germ cell formation in embryos of *Xenopus laevis*. *J. Embryol. Exp. Morphol.* **35**, 617–23.

Ijiri, K. I. (1977). Existence of ultraviolet-labile germ cell determinant in unfertilized eggs of *Xenopus laevis* and its sensitivity. *Develop. Biol.* **55**, 206–11.

Ikenishi, K. & Kotani, M. (1975). Ultrastructure of the 'germinal plasm' in *Xenopus* embryos after cleavage. *Develop. Growth Differ.* **17**, 101–10.

Ikenishi, K., Kotani, M. & Tanabe, K. (1974). Ultrastructural changes associated with UV irradiation in the 'germinal plasm' of *Xenopus laevis*. *Develop. Biol.* **36**, 155–68.

Ikenishi, K. & Nieuwkoop, P. D. (1978). Location and ultrastructure of primordial germ cells (PGCs) in *Ambystoma mexicanum*. *Develop. Growth Differ.* **20**, 1–9.

Ioannou, J. M. (1964). Oogenesis in the guinea pig. *J. Embryol. Exp. Morphol.* **12**, 673–91.

Jarvik, E. (1955). The oldest tetrapods and their forerunners. *Sci. Monthly* **80**, 141–54.

Jarvik, E. (1968*a*). The systematic position of the Dipnoi. In *Current Problems of Lower Vertebrate Phylogeny*, ed. T. Ørvig, pp. 223–45. Wiley-Interscience, New York.

Jarvik, E. (1968*b*). Aspects of vertebrate phylogeny. In *Current Problems of Lower Vertebrate Phylogeny*, ed. T. Ørvig, pp. 497–527. Wiley-Interscience, New York.

Jeon, K. W. & Kennedy, J. R. (1973). The primordial germ cells in early mouse embryos: light and electron microscopic studies. *Develop. Biol.* **31**, 275–84.

Johnston, P. M. (1951). The embryonic history of the germ cells of the largemouth black bass, *Micropterus salmoides salmoides* (Lacépède). *J. Morphol.* **88**, 471–542.

Kalt, M. R. (1973). Ultrastructural observations on the germ line of *Xenopus laevis*. *Z. Zellforsch. Mikrosk. Anat.* **138**, 41–62.

Kalt, M. R., Pinney, H. E. & Graves, K. (1975). Inhibitor induced alterations of chromatoid bodies in male germ line cells of *Xenopus laevis*. *Cell Tissue Res.* **161**, 193–210.

Kamimura, M., Ikenishi, K., Kotani, M. & Matsuno, T. (1976). Observations on the migration and proliferation of gonocytes in *Xenopus laevis*. *J. Embryol. Exp. Morphol.* **36**, 197–207.

Kannankeril, J. V. & Domm, L. V. (1968). Development of the gonads in the female Japanese quail. *Amer. J. Anat.* **123**, 131–46.

Kato, K. (1957). Role of the neural tissue in the differentiation of the notochord from the uninvaginated part of the dorsal blastopore lip of *Triturus* gastrula. *Mem. Coll. Sci. Univ. Kyoto, Ser. B* **24**, 165–70.

Kato, K. (1963). Neuro-notochordal relationship in the development of the explanted pieces taken from the dorsal lip of *Triturus* gastrula. *Mem. Coll. Sci. Univ. Kyoto, Ser. B* **30**, 29–39.

Keller, R. E. (1975). Vital dye mapping of the gastrula and neurula of *Xenopus laevis*. I. Prospective areas and morphogenetic movements of the superficial layer. *Develop. Biol.* **42**, 222–41.

Keller, R. E. (1976). Vital dye mapping of the gastrula and neurula of *Xenopus laevis*. II. Prospective areas and morphogenetic movements of the deeper layer. *Develop. Biol.* **51**, 118–37.

Kennelly, J. J. & Foote, R. H. (1966). Oocytogenesis in rabbits. The role of neogenesis in the formation of the definitive ova and the stability of oocyte DNA measured with tritiated thymidine. *Amer. J. Anat.* **118**, 573–90.

Kessel, R. G. (1971). Cytodifferentiation in the *Rana pipiens* oocyte. II. Intramitochondrial yolk. *Z. Zellforsch. Mikrosk. Anat.* **112**, 313–32.

Knight, A. E. (1963). The embryonic and larval development of the rainbow trout. *Trans. Amer. Fish. Soc.* **92**, 344–55.

Kochav, S. & Eyal-Giladi, H. (1971). Bilateral symmetry in chick embryo: determination by gravity. *Science* **171**, 1027–9.

Kocher-Becker, U. & Tiedemann, H. (1971). Induction of mesodermal and endodermal structures and primordial germ cells in *Triturus* ectoderm by a vegetalizing factor from chick embryos. *Nature, Lond.* **233**, 65–6.

Kocher-Becker, U., Tiedemann, H. & Tiedemann, H. (1965). Exovagination of newt endoderm: cell affinities altered by the mesodermal inducing factor. *Science* **147**, 167–9.

Koebke, J. (1974). Zur Frage der Differenzierungsfähigkeit der Marginalzonebereiche früher Entwicklungsstadien von *Ambystoma mexicanum*. PhD Thesis, University of Cologne, 90pp.

Komar, A. (1969). Primordial germ cells in chick germinal crescent developing as chorio-allantoic or intra-coelomic graft. *Zool. Pol.* **19**, 517–23.

Kotani, M. (1957). On the formation of the primordial germ cells from the presumptive ectoderm of *Triturus* gastrulae. *J. Inst. Polytech. Osaka City Univ.*, *Ser D* **8**, 145–59.

Kotani, M. (1962). The differentiation of Wolffian duct, mesonephros and primordial germ cells in the newt *Triturus pyrrhogaster* after extirpation of the primordium of Wolffian duct. *J. Biol.*, *Osaka City Univ.* **13**, 111–18.

Kupffer, C. (1890). Die Entwicklung von *Petromyzon planeri*. *Arch. Mikrosk. Anat.* **35**, 469–58.

Lacroix, J. C. & Capuron, A. (1966). Localisation et greffe des cellules germinales primordiales chez *Pleurodeles waltlii* Michah. (Amphibien Urodèle). Preuves cytogénétiques. *C. R. Acad. Sci., Paris* **263**, 1244–7.

Leblond, C. P., Steinberger, E. & Roosen–Runge, E. C. (1963). Spermatogenesis. In *Conference on the Physiological Mechanisms Concerned with Conception*, ed. C. G. Hartman, pp. 1–72. Pergamon Press, Oxford.

Leikola, A. (1963). The mesodermal and neural competence of isolated gastrula ectoderm studied by heterogenous inductors. *Ann. Zool. Soc. 'Vanamo'* **25**, 1–50.

Levak-Švajger, B. & Svajger, A. (1970). Differentiation in homografts of isolated germ layers of the rat embryo. *Arch. Sci. Biol. Beograd* **22**, 25–32.

Lewis, W. H. (1942). The formation of the blastodisc in the egg of the zebra fish, *Brachydanio rerio*, illustrated with motion pictures. *Anat. Rec.* **84**, 463–4.

Librera, E. (1964). Effects on gonad differentiation of the removal of vegetal plasm in eggs and embryos of *Discoglossus pictus*. *Acta Embryol. Morphol. Exp.* **7**, 217–223.

Luther, W. (1937). Potenzprüfungen an isolierten Teilstücken der Forellenkeimscheibe. *Wilhelm Roux' Arch. Entwicklungsmech. Organismen* **135**, 359–83.

Luther, W. (1938). Transplantations- und Defektversuche am Organisationszentrum der Forellenkeimscheibe. *Wilhelm Roux' Arch. Entwicklungsmech. Organismen* **137**, 404–34.

Lutz, H. (1955). Contribution expérimentale à l'étude de la formation de l'endoblaste chez les oiseaux. *J. Embryol. Exp. Morphol.* **3**, 59–76.

Lutz, H. & Lutz-Ostertag, Y. L. (1972). Autorégulation du germen chez les oiseaux. *C. R. Soc. Biol.* **166**, 1694–6.

McAlpine, R. J. (1955). Alkaline glycerophosphatase in the developing adrenal, gonads, and reproductive tract of the white rat. *Anat. Rec.* **121**, 407–8.

McCallion, D. J. & Wong, W. T. (1956). A study of the localization and distribution of glycogen in early stages of the chick embryo. *Canad. J. Zool.* **34**, 63–7.

McKay, D. G., Hertig, A. T., Adams, E. C. & Danziger, S. (1953). Histochemical observations on the germ cells of human embryos. *Anat. Rec.* **117**, 201–19.

Mahowald, A. P. & Hennen, S. (1971). Ultrastructure of the 'germ plasm' in eggs and embryos of *Rana pipiens*. *Develop. Biol.* **24**, 37–53.

Marcus, H. (1908). Ueber Mesodermbildung im Gymnophionenkopf. *Sitzungsber. Ges. Morphol. Physiol. München*, 1–11.

Marcus, H. (1938). Ueber die Keimbahn, Keimdrüsen, Fettkörper und Urogenitalverbindung bei *Hypogeophis*. *Biomorph.* **1**, 355–84.

Marin, G. (1959). Gonadogenesi in assenza di gonociti e azione della radiazioni ionizzante sul differenziamento sessuale dell'embrioni di pollo. *Arch. Ital. Anat. Embriol.* **64**, 211–35.

Martin, S. (1959). Recherches descriptives et expérimentales sur les modalités du développement et l'étude de l'évolution du système pronéphrétique chez l'embryon et la larve du Crapaud accoucheur (*Alytes obstetricans* Laur.). PhD Thesis, University of Bordeaux.

Maschkowzeff, A. (1934). Zur Phylogenie der Geshlechtsdrüsen und der Geschlechtsausführgänge bei den Vertebrata auf Grund von Forschungen betreffend die Entwicklung des Mesonephros und der Geschlechtsorgane bei den Acipenseridae, Salmoniden und Amphibien. I. Die Entwicklung des Mesonephros und der Genitaldrüse bei den Acipenseridae und Salmonidae. *Zool. Jahrb.* (*Anat.*) **59**, 1–68.

Maufroid, J. P. & Capuron, A. (1972). Migration des cellules germinales primordiales chez l'amphibien urodele *Pleurodeles waltlii* Michah. *Wilhelm Roux' Arch. Entwicklungsmech. Organismen* **170**, 234–43.

Maufroid, J. P. & Capuron, A. (1973). Mise en évidence expérimentale de cellules germinales dans le mésoderme latéral présumptif de la jeune gastrula de *Pleurodeles waltlii* (Amphibien urodèle). *C. R. Acad. Sci., Paris* **276**, 821–4.

Maufroid, J. P. & Capuron, A. (1977). Induction du mésoderme et des cellules germinales primordiales par l'endoderme chez *Pleurodeles waltlii* (Amphibien, Urodèle): évolution au cours de la gastrulation. *C. R. Acad. Sci., Paris* **284**, 1713–16.

Maufroid, J. P. & Capuron, A. (1978). Recherches récentes sur les cellules germinales primordiales de *Pleurodeles waltlii* (Amphibien Urodèle). *Mém. Soc. Zool. France, Symp. L. Gallien*, **41**, 43–60.

Mendietta, L. (1963). Origine e differenziamento della gonade in '*Gongylus ocellatus*'. *Monit. Zool. Ital.* **70/71**, 201–10.

Merchant-Laros, H. (1976). The role of germ cells in the morphogenesis and cytodifferentiation of the rat ovary. In *Progress in Differentiation Research*, ed. N. Müller-Bérat, pp. 453–62. North-Holland, Amsterdam.

Merchant, H. & Zamboni, L. (1973). Fine morphology of extragonadal germ cells in the mouse. In *The Development and Maturation of the Ovary and its Functions*, ed. H. E. Peters, pp. 95–100. Excerpta Medica, Amsterdam.

Meyer, D. B. (1960). Application of the periodic acid–Schiff technique to whole chick embryos. *Stain Technol.* **35**, 83–9.

Meyer, D. B. (1964). The migration of primordial germ cells in the chick embryo. *Develop. Biol.* **10**, 154–90.

Milaire, J. (1957). Contribution à la connaissance morphologique et cytochimique des bourgeons de membres chez quelques reptiles. *Arch. Biol.* **68**, 429–512.

Mintz, B. (1957a). Embryological development of primordial germ cells in the mouse: influence of a new mutation, W^j. *J. Embryol. Exp. Morphol.* **5**, 396–403.

Mintz, B. (1957b). Germ cell origin and history in the mouse: genetic and histochemical evidence. *Anat. Rec.* **127**, 335–6.

Mintz, B. (1957c). Interaction between two allelic series modifying primordial germ cell development in the mouse embryo. *Anat. Rec.* **128**, 591 (abstr.).

Mintz, B. (1959). Continuity of the female germ cell line from embryo to adult. *Arch. Anat. Microsc. Morphol. Exp.* **48**, 155–72.

Mintz, B. (1960a). Formation and early development of germ cells. In *Symposium on Germ Cells and Development*, pp. 1–24. IIE/A. Baselli, Milan.

Mintz, B. (1960b). Embryological phases of mammalian gametogenesis. *J. Cell. Comp. Physiol.* **56**, 31–48.

Mintz, B. & Russell, E. S. (1955). Developmental modifications of primordial germ cells, induced by *W*-series genes in the mouse embryo. *Anat. Rec.* **122**, 443(abstr.).

Mintz, B. & Russell, E. S. (1957). Gene-induced embryological modifications of primordial germ cells in the mouse. *J. Exp. Zool.* **134**, 207–38.

Modak, S. P. (1966). Analyse expérimentale de l'origine de l'endoblaste embryonnaire chez les oiseaux. *Rev. Suisse Zool.* **73**, 877–908.

Motta, P. & Van Blerkom, J. (1974). Présence d'un matériel caractéristique granulaire dans le cytoplasme de l'ovocyte et dans les premiers stades de la différenciation des cellules embryonnaires. *Bull. Ass. Anat.* **58**, 947–53.

Muchmore, W. B. (1951). Differentiation of the trunk mesoderm in *Ambystoma maculatum. J. Exp. Zool.* **118**, 137–86.

Muchmore, W. B. (1964). Control of muscle differentiation by embryonic neural tissues. *J. Embryol. Exp. Morphol.* **12**, 587–96.

Mukai, H. & Watanabe, H. (1976). Studies in the formation of germ cells in a compound Ascidian *Botryllus primigenus* Oka. *J. Morphol.* **148**, 337–61.

Mulnard, J. (1955). Contribution à la connaissance des enzymes dans l'ontogénèse. Les phosphomonoestérase acide et alcaline dans le développement du rat et de la souris. *Arch. Biol.* **66**, 525–685.

Nakamura, O. (1942). Die Entwicklung der hinteren Körperhälfte bei Urodelen. *Annot. Zool. Jap.* **21**, 169–236.

Nakamura, O. & Matsuzawa, T. (1967). Differentiation capacity of the marginal zone in the morula and blastula of *Triturus pyrrhogaster. Embryologia* **9**, 223–37.

Nakamura, O. & Takasaki, H. (1971). Analysis of causal factors giving rise to the organizer. *Proc. Jap. Acad.* **47**, 499–504.

Nakatsuji, N. (1975). Studies on the gastrulation of amphibian embryos: light and electron microscopic observation of a urodele *Cynops pyrrhogaster. J. Embryol. Exp. Morphol.* **34**, 669–85.

Nedelea, M. & Steopoe, I. (1970). Origine, caractères cytologiques et comportement des gonocytes primaires pendant l'embryogénèse et chez les jeunes larves de *Cyprinus carpio* L. (Téléostéens). *Anat. Anz.* **127**, 338–46.

Nicolet, G. (1970). Analyse autoradiographique de la localisation des différentes ébauches présumptives dans la ligne primitive de l'embryon de poulet. *J. Embryol. Exp. Morphol.* **23**, 79–108.

Nicolet, G. (1971). Avian gastrulation. *Adv. Morphogen.* **9**, 231–62.

Nieuwkoop, P. D. (1947). Experimental investigations on the origin and determination of the germ cells, and on the development of the lateral plates and germ ridges in the urodeles. *Arch. Neer. Zool.* **8**, 1–205.

Nieuwkoop, P. D. (1950). Causal analysis of the early development of the primordial germ cells and the germ ridges in urodeles. *Arch. Anat. Microsc. Morphol. Exp.* **39**, 257–68.

Nieuwkoop, P. D. (1964). Le problème de la lignée germinale chez les urodeles. In *L'origine de la lignée germinale chez les vertébrés et chez quelques groupes d'invertébrés*, ed. E. Wolff, pp. 195–204. Hermann, Paris.

Nieuwkoop, P. D. (1969*a*). The formation of the mesoderm in urodelean amphibians. I. Induction by the endoderm. *Wilhelm Roux' Arch. Entwicklungsmech. Organismen* **162**, 341–73.

Nieuwkoop, P. D. (1969*b*). The formation of the mesoderm in urodelean amphibians. II. The origin of the dorso-ventral polarity of the mesoderm. *Wilhelm Roux' Arch. Entwicklungsmech. Organismen* **163**, 298–315.

Nieuwkoop, P. D. (1970). The formation of the mesoderm in urodelean amphibians. III. The vegetalizing action of the Li ion. *Wilhelm Roux' Arch. Entwicklungsmech. Organismen* **166**, 105–23.

Nieuwkoop, P. D. (1973). The 'organization centre' of the amphibian embryo: its origin, spatial organization and morphogenetic action. *Adv. Morphogen.* **10**, 1–39.

Nieuwkoop, P. D. (1977). Origin and establishment of embryonic polar axes in amphibians. *Curr. Top. Develop. Biol.* **11**, 115–32.

Nieuwkoop, P. D., Boterenbrood, E. C., Kremer, A., Bloemsma, F. F. S. N., Hoessels, E. L. M. J., Meyer, G. & Verheyen, F. J. (1952). Activation and organization of the central nervous system in amphibians. *J. Exp. Zool.* **120**, 1–108.

Nieuwkoop, P. D. & Faber, J. (1975). *Normal Table of Xenopus laevis.* (*Daudin*), 2nd edn, 252pp. North-Holland, Amsterdam.

Nieuwkoop, P. D. & Florschütz, P. A. (1950). Quelques caractères spéciaux de la gastrulation et de la neurulation de l'oeuf de *Xenopus laevis* Daud. et de quelques autres anoures. I. Etude descriptive. *Arch. Biol.* **61**, 113–50.

Nieuwkoop, P. D. & Suminski, E. H. (1959). Does the so-called 'germinal plasm' play an important role in the development of the primordial germ cells. *Arch. Anat. Microsc. Morphol. Exp.* **48**, 189–98.

Nieuwkoop, P. D. & Sutasurya, L. A. (1976). Embryological evidence for a possible polyphyletic origin of the recent amphibians. *J. Embryol. Exp. Morphol.* **35**, 159–67.

Nieuwkoop, P. D. & Ubbels, G. A. (1972). The formation of the mesoderm in urodelean amphibians. IV. Qualitative evidence for the purely 'ectodermal' origin of the entire mesoderm and of the pharyngeal endoderm. *Wilhelm Roux' Arch. Enwicklungsmech. Organismen* **169**, 185–99.

Nieuwkoop, P. D. & Weyer, C. J. (1978). Neural induction, a two-way process. *Med. Biol.*, **56**, 366–371.

Nussbaum, M. (1880). Zur Differenzierung des Geschlechts im Thierreich. *Arch. Mikrosk. Anat.* **18**, 1–121.

Oakberg, E. F. (1955). Degeneration of spermatogonia of the mouse following exposure to X rays, and stages in the mitotic cycle at which cell death occurs. *J. Morphol.* **97**, 39–54.

Oakberg, E. F. (1971). Spermatogonial stem-cell renewal in the mouse. *Anat. Rec.* **169**, 515–31.

Okada, Y. K. & Hama, T. (1943). Examination of regional differences in the inductive activity of the organizer by means of transplantation into ectodermal vesicles. *Proc. Imp. Acad., Tokyo* **19**, 48–53.

Okada, Y. K. & Hama, T. (1944). On the different effects of the amphibian organizer following culture, transplantation and heat treatment. *Proc. Imp. Acad., Tokyo* **20**, 36–40.

Okada, Y. K. & Hama, T. (1945). Prospective fate and inductive capacity of the dorsal lip of the blastopore of the *Triturus* gastrula. *Proc. Jap. Acad.* **21**, 342–8.

Okada, Y. K. & Takaya, H. (1942*a*). Experimental investigation of regional differences in the inductive capacity of the organizer. *Proc. Imp. Acad., Tokyo* **18**, 505–13.

Okada, Y. K. & Takaya, H. (1942*b*). Further studies upon the regional differentiation of the inductive capacity of the organizer. *Proc. Imp. Acad., Tokyo* **18**, 514–19.

Okkelberg, P. (1921). The early history of the germ cells in the brook lamprey, *Entosphenus wilderi* (Goge) up to and including the period of sex differentiation. *J. Morphol.* **35**, 1–127.

Olson, E. C. (1966). Relationships of *Diadectes. Fieldiana, Geol.* **14**, 199–227.

Oppenheimer, J. M. (1936*a*). The development of isolated blastoderms of *Fundulus heteroclitus. J. Exp. Zool.* **72**, 247–69.

Oppenheimer, J. M. (1936*b*). Transplantation experiments on developing teleosts (*Fundulus* and *Perca*). *J. Exp. Zool.* **72**, 409–37.

Oppenheimer, J. M. (1936*c*). Processes of localization in developing *Fundulus. J. Exp. Zool.* **73**, 405–44.

Oppenheimer, J. M. (1938). Potencies of differentiation in the teleostean germ ring. *J. Exp. Zool.* **79**, 185–212.

Oppenheimer, J. M. (1947). Organization of the teleost blastoderm. *Quart. Rev. Biol.* **22**, 105–18.

Oppenheimer, J. M. (1955). The differentiation of derivatives of the lower germ layers in *Fundulus* following the implantation of shield grafts. *J. Exp. Zool.* **128**, 525–60.

Oppenheimer, J. M. (1959*a*). Extraembryonic transplantation of sections of the *Fundulus* embryonic shield *J. Exp. Zool.* **140**, 247–68.

Oppenheimer, J. M. (1959*b*). Extraembryonic transplantation of fragmented shield grafts in *Fundulus*. *J. Exp. Zool.* **142**, 441–60.

Oppenheimer, J. M. (1964). The development of isolated *Fundulus* embryonic shields in salt solution. *Acta Embryol. Morphol. Exp.* **7**, 143–54.

Ortolani, G. (1954). Risultati definitivi sulla distribuzione dei territori presumptivi degli organi, nei germi de Ascidie allo stadio VIII, determini con le marche al carbone. *Pubbl. Staz. Zool. Napoli* **25**, 161–87.

Ortolani, G. (1955). The presumptive territory of the mesoderm in the ascidian egg. *Experientia* **11**, 445.

Ortolani, G. (1958). Cleavage and development of egg fragments in ascidians. *Acta Embryol. Morphol. Exp.* **1**, 247–72.

Ortolani, G. (1959). Ricerche sulla induzione del sistema nervoso nella larve delle Ascidie. *Boll. Zool.* **26**, 341–8.

Ortolani, G. (1971). Sul cell-lineage delle Ascidie. *Boll. Zool.* **38**, 85–8.

Ozdzenski, W. (1967). Observations on the origin of primordial germ cells in the mouse. *Zool. Pol.* **17**, 367–81.

Ozdzenski, W. (1969). Fate of primordial germ cells in the transplanted hind gut of mouse embryos. *J. Embryol. Exp. Morphol.* **22**, 505–10.

Padoa, E. (1963*a*). Le gonade di girini di *Rana esculenta* da uova irradiate con ultravioletto. *Monit. Zool. Ital.* **70/71**, 238–49.

Padoa, E. (1963*b*). Qualche precisazione sulla possibilità di distruggere con l'ultravioletto il plasma germinale dela uova di *Rana esculenta. Boll. Soc. Ital. Biol. Sper.* **40**, 272–5.

Padoa, E. (1964). Il differenziamento sessuale dele gonadi di *Rana esculenta* rese sterili dall'irradiamento con ultravioletto delle uova indivise. *Boll. Zool.* **31**, 811–25.

Pala, M. (1970). The embryonic history of the primordial germ cells in *Gambusia holbrookii* (Grd). *Boll. Zool.* **37**, 49–62.

Parker, H. W. (1956). Viviparous coecileans and amphibian phylogeny. *Nature, Lond.* **178**, 250–2.

Parsons, T. S. & Williams, E. E. (1963). The relationships of the modern Amphibia: a reexamination. *Quart. Rev. Biol.* **38**, 26–53.

Pasteels, J. (1936). Etudes sur la gastrulation des vertébrés méroblastiques. I. Téléostéens. *Arch. Biol.* **47**, 205–308.

Pasteels, J. (1937*a*). Sur l'origine de la symétrie bilatérale des amphibiens anoures. *Arch. anat. Microsc. Morphol. Exp.* **33**, 279–300.

Pasteels, J. (1937*b*). Etudes sur la gastrulation des vertébrés méroblastiques. II. Reptiles. *Arch. Biol.* **48**, 107–84.

Pasteels, J. (1937*c*). Etudes sur la gastrulation des vertébrés méroblastiques. III. Oiseaux. IV. Conclusions générales. *Arch. Biol.* **48**, 381–488.

Pasteels, J. (1940). Un aperçu comparatif de la gastrulation chez les chordés. *Biol. Rev.* **15**, 59–106.

Pasteels, J. (1942). New observations concerning the maps of presumptive areas of the young amphibian gastrula (*Amblystoma* and *Discoglossus*). *J. Exp. Zool.* **89**, 255–81.

Pasteels, J. (1953). Contribution à l'étude du développement des reptiles. I. Origine et migration des gonocytes chez deux Lacertiliens (*Mabuia megalura* et *Chamaeleo bitaeniatus*). *Arch. Biol.* **64**, 227–45.

Pasteels, J. (1957*a*). Une table analytique du développement des reptiles. I. Stades de gastrulation chez les Chéloniens et les Lacertiliens. *Ann. Soc. Roy. Zool. Belg.* **87**, 217–41.

Pasteels, J. (1957*b*). La formation de l'endophylle et de l'endoblaste vitellin chez les reptiles, Chéloniens et Lacertiliens. *Acta Anat.* **30**, 601–12.

Pasteels, J. (1964). La lignée germinale chez les reptiles et chez les mammifères. In *L'origine de la lignée germinale chez les vertébrés et chez quelques groupes d'invertébrés*, ed. E. Wolff, pp. 265–80. Hermann, Paris.

Pelliniemi, L. J. (1975). Ultrastructure of gonadal ridge in male and female pig embryos. *Anat. Embryol.* **147**, 19–34.

Pelliniemi, L. J. (1976). Human primordial germ cells during migration and entrance to the gonadal ridge. *J. Cell Biol.* **78**, 226 (abstr.).

Peter, K. (1938). Untersuchungen über die Entwicklung des Dotterentoderm. II. Die Entwicklung des Entoderms bei Reptilien. *Z. Mikrosk. Anat. Forsch.* **44**, 498–531.

Peters, H. (1970). Migration of gonocytes into the mammalian gonad and their differentiation. *Phil. Trans. Roy. Soc., Ser. B.* **259**, 91–101.

Peters, H. & Crone, M. (1967). DNA synthesis in oocytes of mammals. *Arch. Anat. Microsc. Morphol. Exp.* **56**, 160–70.

Reagan, F. P. (1916). Some results and possibilities of early embryonic castration. *Anat. Rec.* **11**, 251–67.

Reed, C. A. (1960). Polyphyletic or monophyletic ancestry of mammals, or what is a class? *Evolution* **14**, 314–22.

Reed, S. C. & Stanley, H. P. (1972). Fine structure of spermatogenesis in the South African clawed toad *Xenopus laevis* Daudin. *J. Ultrastruct. Res.* **41**, 277–95.

Reinhard, L. (1924). Die Entwicklung des Parablasts und seine Bedeutung bei Teleostiern nebst der Frage über die Entstehung der Urgeschlechtszellen. *Arch. Mikrosk. Anat. Entwicklungsmech.* **103**, 339–56.

Reverberi, G. (1961). L'uovo delle ascidie e le sue potenze organo-formative. *Rend. Ist. Sci. Univ. Camerino* **2**, 187–209.

Reverberi, G. (1971*a*). Ascidians. In *Experimental Embryology of Marine and Freshwater Inverterates*, ed. G. Reverberi, pp. 507–50. North-Holland, Amsterdam.

Reverberi, G. (1971*b*). Amphioxus. In *Experimental Embryology of Marine and Freshwater Invertebrates*, ed. G. Reverberi, pp. 551–72. North-Holland, Amsterdam.

Reverberi, G. & Gorgone, I. (1962). Gigantic tadpoles from ascidian eggs fused at the 8-cell stage. *Acta Embryol. Morphol. Exp.* **5**, 104–12.

Reverberi, G. & Ortolani, G. (1962). Twin larvae from halves of the same egg in ascidians. *Develop. Biol.* **5**, 84–100.

Reynaud, G. (1968). Contribution à l'étude des rapports entre soma et germen chez les oiseaux au moyen d'une technique d'injection des gonocytes primordiaux. Dr. Sci. Thesis, University of D'aix-Marseille, 72pp.

Reynaud, G. (1969). Transfert de cellules germinales primordiales de dindon à l'embryon de poulet par injection intravasculaire. *J. Embryol. Exp. Morphol.* **21**, 485–507.

Reynaud, G. (1970*a*). Transfert de cellules germinales primordiales chez les oiseaux au moyen d'une technique d'injection intravasculaire. *Ann. Fac. Sci. Marseille*, **43**, 101–5.

Reynaud, G. (1970*b*). Transferts interspécifiques de gonocytes primordiaux chez les oiseaux. *Ann. Biol.* **9**, 495–500.

Reynaud, G. (1971*a*). Etude comparée de la multiplication des cellules germinales à la fin de la première semaine de la vie embryonnaire chex trois espèces de Gallinacés

(*Gallus domesticus, Meleagris gallopavo, Cortunix coturnix japonica*). *Experientia* **27**, 427–8.

Reynaud, G. (1971*b*). Sur les durées respectives du pouvoir attractif de l'épithélium germinatif et de la réactivité des cellules germinales à ce pouvoir, chez l'embryon de poulet. *Experientia* **27**, 940–1.

Reynaud, G. (1976*a*). Capacités reproductrices et descendance de poulets ayant subi un transfert de cellules germinales primordiales durant la vie embryonnaire. *Wilhelm Roux' Arch. Develop. Biol.* **179**, 85–110.

Reynaud, G. (1976*b*). Etude de la localisation des cellules germinales primordiales chez l'embryon de caille japonaise au moyen d'une technique d'irradiation aux rayons ultraviolets. *C. R. Acad. Sci., Paris* **282**, 1195–8.

Richards, A. & Thompson, J. T. (1921). The migration of the primary sex-cells of *Fundulus heteroclitus*. *Biol. Bull.* **40**, 325–49.

Risley, P. L. (1933). Contributions on the development of the reproductive system in *Sternotherus odoratus* (Latreille). I. The embryonic origin and migration of the primordial germ cells. *Z. Zellforsch. Mikrosk. Anat.* **18**, 459–92.

Rogulska, T. (1968). Primordial germ cells in normal and transected duck blastoderms. *J. Embryol. Exp. Morphol.* **20**, 247–60.

Rogulska, T. (1969). Migration of the chick primordial germ cells from the intra-coelomically transplanted germinal crescent into the genital ridges. *Experientia* **25**, 631–2.

Rogulska, T., Ozdzenski, W. & Komar, A. (1971). Behaviour of mouse primordial germ cells in the chick embryo. *J. Embryol. Exp. Morphol.* **25**, 155–64.

Romer, A. S. (1947). Review of the Labyrinthodontia. *Bull. Mus. Comp. Zool.* **99**, 1–368.

Romer, A. S. (1958). Tetrapod limbs and early tetrapod life. *Evolution* **12**, 365–9.

Romer, A. S. (1966). Early reptilian evolution reviewed. *Evolution* **21**, 821–33.

Romer, A. S. (1971). Unorthodoxies in reptilian phylogeny. *Evolution* **25**, 103–12.

Roosen-Runge, E. C. (1962). The process of spermatogenesis in mammals. *Biol. Rev.* **37**, 343–77.

Roosen-Runge, E. C. (1977). *The Process of Spermatogenesis in Animals*, 212pp. Cambridge University Press.

Roosen-Runge, E. C. & Leik, J. (1968). Gonocyte degeneration in the postnatal male rat. *Amer. J. Anat.* **122**, 275–300.

Rose, S. M. (1939). Embryonic induction in the ascidia. *Biol. Bull.* **77**, 216–32.

Rosenquist, G. C. (1966). A radioautographic study of labeled grafts in the chick blastoderm. Development from primitive streak stages to stage 12. *Carnegie Inst. Wash. Publ.* **38**, 71–110.

Rosenquist, G. C. (1970). The origin and movement of nephrogenic cells in the chick embryo as determined by radio-autographic mapping. *J. Embryol. Exp. Morphol.* **24**, 367–80.

Rosenquist, G. C. (1971). The location of the pregut endoderm in the chick embryo at the primitive streak stage as determined by radio-autographic mapping. *Develop. Biol.* **26**, 323–35.

Rudkin, G. T. & Griech, H. A. (1962). On the persistence of oocyte nuclei from fetus to maturity in the laboratory mouse. *J. Cell Biol.* **12**, 169–76.

Runnström, J. (1967). The mechanism of control of differentiation in early development of the sea urchin. A tentative discussion. *Exp. Biol. Med.* **1**, 52–62.

Satoh, N. (1974). An ultrastructural study of sex differentiation in the teleost, *Oryzias latipes*. *J. Embryol. Exp. Morphol.* **32**, 195–215.

Sawano, J. (1959). Primordial germ cells in human embryos. *Hiroshima Med. J.* **12**, 954–60.

Saxén, L. & Toivonen, S. (1962). *Primary Embryonic Induction*, 271pp. Logos Press, London.

Schaeffer, B. (1968). The origin and basic radiation of the Osteichthyes. In *Current Problems of Lower Vertebrate Phylogeny*, ed. T. Ørvig, pp. 207–22. Wiley-Interscience, New York.

Schechtmann, A. M. (1934). Unipolar ingression in *Triturus torosus*: a hitherto undescribed movement in the pregastrular stages of a urodele. *Univ. Calif. Publ. Zool.* **39**, 303–10.

Schmalhausen, I. I. (1968). *The Origin of Terrestrial Vertebrates*. Academic Press, New York & London.

Schreiner, K. E. (1955). Studies on the gonad of *Myxine glutinosa* L. *Univ. Bergen Årb Naturv. R.* **8**, 1–36.

Semenova-Tian-Shanskaya, A. G. (1969). Primordial germ cells during migration to gonad anlage in human embryos. *Arch. Anat. Histol. Embryol.* **56**, 3–8.

Simkins, C. S. (1923). On the origin and migration of the so-called primordial germ cells in the mouse and the rat. *Acta Zool., Stockholm* **4**, 241–78.

Simkins, C. S. & Asana, J. J. (1931). Development of the sex glands of *Calotes*. I. Cytology and growth of the gonads prior to hatching. *Quart. J. Microsc. Sci.* **74**, 133–49.

Simon, D. (1957). La migration des cellules germinales de l'embryon de poulet vers les ébauches gonadiques; preuves expérimentales. *C. R. Soc. Biol.* **151**, 1576.

Simon, D. (1960). Contribution à l'étude de la circulation et du transport des gonocytes primaires dans les blastodermes d'oiseaux cultivés *in vitro*. *Arch. Anat. Microsc. Morphol. Exp.* **49**, 93–176.

Simon, D. (1961). Association de blastodermes d'oiseaux en culture *in vitro*. Application de cette méthode à la migration des gonocytes primaires d'un embryon à un autre embryon. *Colloq. Int. Centre Nat. Rech. Sci.* **101**, 269–75.

Simon, D. (1964). La lignée germinale des oiseaux et la migration des gonocytes primaires. In *L'origine de la lignée germinale chez les vertébrés et chez quelques groupes d'invertébrés*, ed. E. Wolff, pp. 237–62. Hermann, Paris.

Simone-Santoro, I. de (1967). Ultrastruttura della cellula germinale primordiale e dell'oogonio in embrione di pollo. *Boll. Soc. Ital. Biol. Sper.* **43**, 906–7.

Simpson, G. G. (1959). Mesozoic mammals and the polyphyletic origin of mammals. *Evolution* **13**, 405–14.

Singh, R. P. & Meyer, D. B. (1967). Primordial germ cells in blood smears from chick embryos. *Science* **156**, 1503–4.

Skreb, N., Švajger, A. & Levak-Švajger, B. (1976). Developmental potentialities of the germ layers in mammals. In *Embryogenesis in Mammals*, ed. K. Elliott & M. O'Connor, *Ciba Foundation Symposium* **40**, pp. 27–45. Elsevier, Amsterdam.

Smith, L. D. (1964). A test of the capacity of presumptive somatic cells to transfer into primordial germ cells in the Mexican axolotl. *J. Exp. Zool.* **156**, 229–42.

Smith, L. D. (1965). Transplantation of nuclei of primordial germ cells into enucleated eggs of *Rana pipiens*. *Proc. Nat. Acad. Sci. USA* **54**, 101–7.

Smith, L. D. (1966). The role of a 'germinal plasm' in the formation of primordial germ cells in *Rana pipiens*. *Develop. Biol.* **14**, 330–47.

Smith, L. D. (1975). Molecular events during oocyte maturation. In *The Biochemistry of Animal Development*, ed. R. Weber, vol. 3, *Molecular Aspects of Animal Development*, pp. 1–42. Academic Press, New York & London.

Smith, L. D. & Ecker, R. E. (1970). Regulatory processes in the maturation and early cleavage of amphibian eggs. *Curr. Top. Develop. Biol.* **5**, 1–38.

Smith, L. D. & Williams, M. A. (1975). Germinal plasm and determination of the primordial germ cells. In *The Developmental Biology of Reproduction*, ed. C. L. Markert, pp. 3–24. Academic Press, New York & London.

Spiegelman, M. & Bennett, D. (1973). A light- and electron-microscopic study of primordial germ cells in the early mouse embryo. *J. Embryol. Exp. Morphol.* **30**, 97–118.

Spratt, N. T. (1946). Formation of the primitive streak in the explanted chick blastoderm marked with carbon particles. *J. Exp. Zool.* **103**, 259–304.

Spratt, N. T. (1952). Localisation of the prospective neural plate in the early chick blastoderm. *J. Exp. Zool.* **120**, 109–30.

Spratt, N. T. (1955). Analysis of the organizer center in the early chick embryo. I. Localisation of prospective notochord and somite cells. *J. Exp. Zool.* **128**, 121–63.

Spratt, N. T. (1957a). Analysis of the organizer center in the early chick embryo. II. Studies on the mechanics of notochord elongation and somite formation. *J. Exp. Zool.* **134**, 577–612.

Spratt, N. T. (1957b). Analysis of the organizer center in the early chick embryo. III. Regulative properties of the chorda and somite centers. *J. Exp. Zool.* **135**, 319–53.

Spratt, N. T. & Haas, H. (1960a). Morphogenetic movement in the lower surface of the unincubated and early chick blastoderm. *J. Exp. Zool.* **144**, 139–57.

Spratt, N. T. & Haas, H. (1960b). Importance of morphogenetic movements in the lower surface of the young chick blastoderm. *J. Exp. Zool.* **144**, 257–75.

Spratt, N. T. & Haas, H. (1961). Integrative mechanisms in development of the early chick blastoderm. II. Role of morphogenetic movements and regenerative growth in synthetic and topographically disarranged blastoderms. *J. Exp. Zool.* **147**, 57–93.

Spratt, N. T. & Haas, H. (1965). Germ layer formation and the role of the primitive streak in the chick. I. Basic architecture and morphogenetic tissue movements. *J. Exp. Zool.* **158**, 9–38.

Stahl, B. J. (1974). *Vertebrate History: Problems in Evolution*, 594pp. McGraw-Hill, New York.

Stärk, O. J. (1955). Entwicklung der Gonaden und Geschlechtszellen bei *Triton alpestris, cristatus* und *taeniatus*, mit besonderer Berüchsichtigung ihrer Verscheidenheiten. *Z. Zellforsch. Mikrosk. Anat.* **41**, 285–334.

Stensiö, E. (1968). The cyclostomes with special reference to the diphyletic origin of the Petromyzontidae and Myxinidae. In *Current Problems of Lower Vertebrate Phylogeny*, ed. T. Ørvig, pp. 13–71. Wiley-Interscience, New York.

Stolk, A. (1958). Extra-regional oocytes in teleosts. *Nature, Lond.* **182**, 1241.

Sud, B. N. (1961). Morphological and histochemical studies of the chromatoid body and related elements in the spermatogenesis of the rat. *Quart. J. Microsc. Sci.* **102**, 495–505.

Sudarwati, S. & Nieuwkoop, P. D. (1971). Mesoderm formation in the anuran *Xenopus laevis* (Daudin). *Wilhelm Roux' Arch. Entwicklungsmech. Organismen* **166**, 189–204.

Sutasurya, L. A. & Nieuwkoop, P. D. (1974). The induction of the primordial germ cells in the urodeles. *Wilhelm Roux' Arch. Entwicklungsmech. Organismen* **175**, 199–220.

Swartz, W. J. (1975). Effect of steroids on definitive localization of primordial germ cells in the chick embryo. *Amer. J. Anat.* **142**, 499–513.

Swartz, W. J. & Domm, L. V. (1972). A study on division of primordial germ cells in the early chick embryo. *Amer. J. Anat.* **135**, 51–70.

Swift, C. H. (1914). Origin and early history of the primordial germ cells in the chick. *Amer. J. Anat.* **15**, 483–516.

Szarski, H. (1962). The origin of the amphibia. *Quart. Rev. Biol.* **37**, 189–241.

Szarski, H. (1968). The origin of vertebrate foetal membranes. *Evolution* **22**, 211–14.

Takamoto, K. (1953). The development of entoderm free embryo. *J. Inst. Polytech. Osaka City Univ., Ser. D.* **4**, 51–60.

Takasaki, H. & Yagura, M. (1975). Remarks on dorso-ventral gradient in the mesoderm formation in *Xenopus laevis. Bull. Osaka Kyoiku Univ.* **24**, 145–58.

Tanabe, K. & Kotani, M. (1974). Relationship between the amount of the 'germinal plasm' and the number of primordial germ cells in *Xenopus laevis. J. Embryol. Exp. Morphol.* **31**, 89–98.

Thomson, K. S. (1968). A critical review of the diphyletic theory of rhipidistian–amphibian relationships. In *Current Problems of Lower Vertebrate Phylogeny*, ed. T. Ørvig, pp. 285–306. Wiley-Interscience, New York.

Toivonen, S. (1953). Bone-marrow of the guinea-pig as a mesodermal inductor in implantation experiments with embryos of *Triturus. J. Embryol. Exp. Morphol.* **1**, 97–104.

Tribe, M. & Brambell, F. W. R. (1932). The origin and migration of the primordial germ cells of *Sphenodon punctatus. Quart. J. Microsc. Sci.* **75**, 251–82.

Tung, T. C., Chang, C. Y. & Tung, Y. F. Y. (1945). Experiments on the developmental potencies of blastoderms and fragments of teleostean eggs separated latitudinally. *Proc. Zool. Soc. Lond.* **115**, 175–88.

Tung, T. C., Wu, S. C. & Tung, Y. F. Y. (1958). The development of isolated blastomeres of amphioxus. *Acta Biol. Exp. Sinica* **6**, 57–90.

Tung, T. C., Wu, S. C. & Tung, Y. F. Y. (1959). The developmental potencies of the blastomere layers in amphioxus egg at the 32-cell stage. *Acta Biol. Exp. Sinica* **6**, 191–210.

Tung, T. C., Wu, S. C. & Tung, Y. F. Y. (1960a). The developmental potencies of the blastomere layers in amphioxus egg at the 32-cell stage. *Sci. Sinica* **9**, 119–41.

Tung, T. C., Wu, S. C. & Tung, Y. F. Y. (1960b). Rotation of the animal blastomeres in amphioxus egg at the 8-cell stage. *Sci. Rec.* **4**, 391–4.

Tung, T. C., Wu, S. C. & Tung, Y. F. Y. (1961). Differentiation of the prospective ectodermal and endodermal cells after transplantation to new surroundings in amphioxus. *Acta Biol. Exp. Sinica* **7**, 253–61.

Tung, T. C., Wu, S. C. & Tung, Y. F. Y. (1962a). The presumptive areas of the egg of amphioxus. *Sci. Sinica* **11**, 629–44.

Tung, T. C., Wu, S. C. & Tung, Y. F. Y. (1962b). Experimental studies on the neural induction in amphioxus. *Sci. Sinica* **11**, 805–20.

Tung, T. C., Wu, S. C., Tung, Y. F. Y., Yan, S. Y. & Tu, M. (1965). Development of amphioxus egg studied by combination of the animal and vegetal blastomeres at the 8- and 16-cell stages. *Acta Biol. Exp. Sinica* **10**, 318–31.

Tung, Y. F. Y., Luh, T. Y. & Tung, S. M. (1965). Experimental studies on the regulation capacities of endoderm in amphioxus. *Acta Biol. Exp. Sinica* **10**, 332–45.

Ubbels, G. A. (1977). Symmetrisation of the fertilized egg of *Xenopus laevis. Mém. Soc. Zool. France*, **41**, *Symp. L. Gallien*, 103–16.

Ubisch, L. von (1951). Gibt es Organisatoren oder Induktoren im Ascidienkeim? *Acta Biotheor.* **9**, 185–96.

Ubisch, L. von (1952). Die Entwicklung der Monascidien. *Verh. K.N. A.W. Natuurk.*, 2nd ser. **49**, 1–56.

Ubisch, L. von (1963). Über Induktion bei Amphioxus und Ascidien. *Wilhelm Roux' Arch. Entwicklungsmech. Organismen* **154**, 466–94.

Ukeshima, A. & Fujimoto, T. (1975). Observations on the migration and distribution of the chick primordial germ cells by application of the PAS reaction to whole embryos. *Acta. Anat. Nippon.* **50**, 15–21.

Vakaet, L. (1950). La répartition des acides nucléiques au cours du grand accroissement de l'oeuf de *Lebistes reticulatus. Bull. Acad. Roy. Belg. Cl. Sci.* **36**, 941–6.

Vakaet, L. (1955). Recherches cytologiques et cytochimiques sur l'organisation de l'oocyte I de *Lebistes reticulatus. Arch. Biol.* **66**, 1–73.

Vakaet, L. (1962). Some new data concerning the formation of the definitive endoblast in the chick embryo. *J. Embryol. exp. Morphol.* **10**, 38–57.

Vakaet, L. (1964). Diversité fonctionnelle de la ligne primitive du blastoderme de poulet. *C. R. Soc. Biol.* **158**, 1964.

Vakaet, L. (1965). Résultats de la greffe de noeuds de Hensen d'âge différent sur le blastoderme de poulet. *C. R. Soc. Biol.* **159**, 232.

Vakaet, L. (1970). Cinephotomicrographic investigations of gastrulation in the chick blastoderm. *Arch. Biol.* **81**, 387–426.

Van Limborgh, J. (1957). De ontwikkeling van de asymmetrie der gonaden by het embryo van de eend. PhD Thesis, University of Utretcht, The Netherlands, 203pp.

Van Limborgh, J. (1958). Number and distribution of the primary germ cells in duck embryos in the 28- to 36-somite stages. *Acta Morphol. Neerl.-Scand.* **2**, 119–33.

Van Limborgh, J. (1960). Number and distribution of the primary germ cells in duck embryos in the 37- to 44-somite stages. *Acta Morphol. Neerl.-Scand.* **3**, 263–82.

Van Limborgh, J. (1961). The origin of the asymmetrical distribution of the primary germ cells in late somite stage duck embryos. *Acta Morphol. Neerl.-Scand.* **4**, 261–80.

Van Limborgh, J. (1968). Number and distribution of the primary germ cells in early postsomite state duck embryos. *Acta Morphol. Neerl.-Scand.* **7**, 117–44.

Van Limborgh, J., Van Deth, J. H. M. & Tacoma, J. (1960). The early gonadal capillary systems of duck embryos: structure, and velocity of the blood. *Acta Morphol. Neerl.-Scand.* **3**, 35–47.

Vandebroek, G. (1936). Les mouvements morphogénétiques au cours de la gastrulation chez *Scyllium canicula* Cuv. *Arch. Biol.* **47**, 499–584.

Vandebroek, G. (1938). Inductieverschijnselen in de ontwikkeling van de Ascidienkiem. *Natuurwetensch. Tijdschr.* **20**, 234–9.

Vandebroek, G. (1961). On some aspects of the organization of the ascidian embryo. In *Symposium on Germ Cells and Earliest Stages of Development*, pp. 277–82. IIE/A. Baselli, Milan.

Vanneman, A. S. (1917). The early history of the germ cells in the armadillo, *Tatusia novemcincta. Amer. J. Anat.* **22**, 341–63.

Vannini, E. (1962). Esperimenti sul differenziamento regionale della gonade in embrioni e larve di *Bufo. Atti. Accad. Sci. Ist. Bologna, Ser. 11* **9**, 30–6.

Vannini, E. & Gardenghi, G. (1964). Esperimenti di asportazione quasi totale dell'abbozzo genitale presuntivo in embrioni di *Bufo bufo. Boll. Zool.* **31**, 41–53.

Vannini, E. & Giorgi, P. P. (1969). Organogenesi dell'apparato urogenitale degli anfibi: agenesi ed interruzione del dotto di Wolff in embrioni di *Bufo bufo. Arch. Ital. Anat. Embriol.* **74**, 111–43.

Vivien, J. H. (1964). Origine de la lignée germinale chez les poissons. In *L'origine de la lignée germinale chez les vertébrés et chez quelques groupes d'invertébrés*, ed. E. Wolff, pp. 281–308. Hermann, Paris.

Vogt, W. (1929). Gestaltungsanalyse am Amphibienkeim mit örtlicher Vitalfärbung. II. Gastrulation und Mesodermbildung bei Urodelen und Anuren. *Wilhelm Roux' Arch. Entwicklungsmech. Organismen* **120**, 384–706.

Waddington, C. H. (1932). Experiments on the development of chick and duck embryos cultivated *in vitro*. *Phil. Trans. Roy. Soc., Ser. B* **221**, 179–230.

Wakahara, M. (1977). Partial characterization of 'primordial germ cell-forming activity' localized in vegetal pole cytoplasm in anuran egg. *J. Embryol. Exp. Morphol.* **39**, 221–33.

Wake, M. H. (1968). Evolutionary morphology of the coecilian urogenital system. I. The gonads and the fat bodies. *J. Morphol.* **126**, 291–332.

Waldeyer, W. (1870). *Eierstock und Ei. Ein Beitrag zur Anatomie und Entwicklungsgeschichte der Sexualorgane*, 174pp. W. Engelmann, Leipzig.

Wallace, H., Morray, J. & Langridge, W. H. R. (1971). Alternative model for gene amplification. *Nature New Biol.* **230**, 201–3.

Wartenburg, H., Holstein, A. F. & Vossmeyer, J. (1971). Zur Cytologie der pränatalen Gonaden-Entwicklung beim Menschen. II. Elektronenmikroskopischen Untersuchungen über die Cytogenese von Gonocyten und fetalen Spermatogonien im Hoden. *Z. Anat. Entwicklungsgesch.* **134**, 165–85.

Watson, D. M. S. (1940). The origin of the frogs. *Phil. Trans. Roy. Soc., Edinb.* **60**, 195–231.

Watson, D. M. S. (1954). On *Bolosaurus* and the origin and classification of reptiles. *Bull. Mus. Comp. Zool.* **111**, 299–450.

Webb, A. C. (1976). An autoradiographic study of tritiated uridine incorporation into the larval ovary of *Xenopus laevis*. *Anat. Rec.* **184**, 285–300.

Weismann, A. (1883). *Die Entstehung der Sexualzellen bei den Hydromedusen, zugleich ein Beitrag zur Kenntniss des Baues und der Lebenserscheinungen dieser Gruppe*, 296pp. Fischer, Jena.

Weismann, A. (1885). *Die Continuität des Keimplasmas als Grundlage einer Theorie der Vererbung*, 122pp. Fischer, Jena.

Weismann, A. (1892). *Das Keimplasma. Eine Theorie der Vererbung*, 628pp. Fischer, Jena.

Weissenberg, R. (1933). Gastrulation und Urdarmdifferenzierung beim Neunauge im Vergleich mit Ergebnissen von Vogt's Gestaltungsanalyse am Amphibienkeim. *Sitzungsber. Ges. Naturforsch. Freunde Berlin* **8/10**, 388–417.

Westoll, T. S. (1943). The origin of the tetrapods. *Biol. Rev.* **18**, 78–98.

Weyer, C. J., Nieuwkoop, P. D. & Lindenmayer, A. (1977). A diffusion model for mesoderm induction in amphibian embryos. *Acta Biotheor.* **26**, 164–80.

Whitington, P. McD. & Dixon, K. E. (1975). Quantitative studies of germ plasm and germ cells during early embryogenesis of *Xenopus laevis*. *J. Embryol. Exp. Morphol.* **33**, 57–74.

Wickstead, J. H. (1975). Chordata: Acrania (Cephalochordata). In *Reproduction of Marine Invertebrates*, ed. A. C. Giese & J. S. Pearse, vol. 2, pp. 283–319. Academic Press, New York & London.

Williams, E. E. (1959). Gadow's Arcualia and the development of tetrapod vertebrae. *Quart. Rev. Biol.* **34**, 1–32.

Williams, M. A. & Smith, L. D. (1971). Ultrastructure of the 'germinal plasm' during maturation and early cleavage in *Rana pipiens*. *Develop. Biol.* **25**, 568–80.

Willier, B. H. (1937). Experimentally produced sterile gonads and the problem of the origin of germ cells in the chick embryo. *Anat. Rec.* **70**, 89–112.

Witschi, E. (1935). Origin of asymmetry in the reproductive system of birds. *Amer. J. Anat.* **56**, 119–41.

Witschi, E. (1948). Migration of the germ cells of human embryos from the yolk sac to the primitive gonadal folds. *Carnegie Inst. Wash., Contrib. Embryol.* **32**, 67–80.

Witschi, E. (1951). Embryogenesis of the adrenal and the reproductive glands. *Recent Progr. Hormone Res.* **6**, 1–27.

Witschi, E. (1957). The inductor theory of sex differentiation. *J. Fac. Sci. Hokkaido Univ., Ser. VI Zool.* **18**, 428–39.

Witschi, E. (1971). Mechanisms of sexual differentiation: experiments with *Xenopus laevis*. In *Hormones in Development*, ed. M. Hamburgh & E. J. W. Barrington, pp. 601–18. Meredith Corporation, New York.

Wolf, L. E. (1931). The history of the germ cells in the viviparous teleost *Platypaecilus maculatus*. *J. Morphol.* **52**, 115–53.

Wolk, M. & Eyal-Giladi, H. (1977). The dynamics of antigenic changes in the epiblast and hypoblast of the chick during the process of hypoblast, primitive streak and head process formation, as revealed by immunofluorescence. *Develop. Biol.* **55**, 33–45.

Wylie, C. C., Bancroft, M. & Haesman, J. (1976). The formation of the gonadal ridge in *Xenopus laevis*. II. A scanning electron microscope study. *J. Embryol. Exp. Morphol.* **35**, 139–48.

Wylie, C. C. & Haesman, J. (1976). The formation of the gonadal ridge in *Xenopus laevis*. I. A light and transmission electron microscope study. *J. Embryol. Exp. Morphol.* **35**, 125–38.

Wylie, C. C. & Roos, T. B. (1976). The formation of the gonadal ridge in *Xenopus laevis*. III. The behaviour of isolated primordial germ cells *in vitro*. *J. Embryol. Exp. Morphol.* **35**, 149–57.

Yamada, T. (1938). Der Determinationszustand des Rumpfmesoderms von Molchkeimen nach der Gastrulation. *Wilhelm Roux' Arch. Entwicklungsmeh. Organismen* **137**, 151–270.

Yamada, T. (1939). Wechselseitige Induktion zwischen Medullaranlage und Ursegmentmaterial des Molchkeimes, dargestellt an zusammengesetzten Isolaten. *Okajimas Folia Anat. Jap.* **18**, 569–71.

Yamada, T. (1940). Beeinflussung der Differenzierungsleistung des isolierten Mesoderms von Molchkeimen durch zugefügtes Chorda- und Neuralmaterial. *Okajimas Folia Anat. Jap.* **19**, 131–97.

Zamboni, L. & Merchant, H. (1973). The fine morphology of mouse primordial germ cells in extragonadal locations. *Amer. J. Anat.* **137**, 299–336.

Ziegler, H. E. & Ziegler, F. (1892). Beiträge zur Entwicklungsgeschichte von *Torpedo*. *Arch. Mikrosk. Anat.* **39**, 56–102.

Züst, B. & Dixon, K. E. (1975). The effect of UV irradiation of the vegetal pole of *Xenopus laevis* eggs on the pesumptive primordial germ cells. *J. Embryol. Exp. Morphol.* **34**, 209–20.

Züst, B. & Dixon, K. E. (1977). Events in the germ cell lineage after entry of the primordial germ cells into the genital ridges in normal and UV-irradiated *Xenopus laevis*. *J. Embryol. Exp. Morphol.* **41**, 33–46.

Author index

References to third and following authors are given in italics.

Abramowicz, 87
Adams, 64; *65*, *125* (McKay *et al.*)
Aisenstadt, 92
Albert, 18
Allen, 93, 101, 118
Amanuma, 88, 89
Ancel, 10, 18
Arkenberg, 11
Asana, 111
Asayama, 65, 66, 88, 89, 100
Aubry, 77, 83, 105; *85* (Bounoure *et al.*)
Avery, 101

Baer, von, 129
Baker, 6, 75
Balfour, 25
Balinsky, 18, 57
Ballard, 24, 27, 28, 29, 31
Bancroft, 104; *105*, *115* (Wylie *et al.*)
Beams, 56, 57, 59, 82
Beard, 94
Beaumont, 74
Belsare, 107, 108, 118
Bennett, 68, 79, 100, 123, 124, 125
Benoit, 78
Bergeaud, 110
Berrill, 39, 96, 108
Bertmar, 132
Blackler, 57, 77, 78, 82, 83, 91, 105, 107, 115, 146
Blandau, 99, 125
Blocker, 109, 120
Bloemsma, *17* (Nieuwkoop *et al.*)
Borum, 75
Boterenbrood, ix, 13, 14, 18, 91; *17* (Nieuw-koop *et al.*)
Bounoure, ix, 1, 5, 54, 56, 71, 77, 81, 82, 83, 85, 105, 115
Bouvet, 27
Boveri, 96, 108
Brambell, 71, 77, 82, 99, 101, 111, 125
Brauer, 22
Bruel, 72, 97

Bruel-Beaudemon, 100, 101
Buehr, 77, 83

Cambar, 1, 2, 57, 82, 106, 107, 115
Capuron, 72, 88, 89, 90, 91, 104, 106, 116, 117
Carroll, 136
Castro-Correia, 45
Celestino da Costa, 54, 100, 123
Chang, *31* (Tung *et al.*)
Chieffi, 107
Chiquoine, 62, 64, 65, 99, 123, 125
Chrétien, 65, 75, 100, 123
Clark, 65, 68, 100, 124
Clavert, 42
Clawson, 64, 72, 97, 122
Clayton, 84
Clemens, 139
Clermont, 76
Coggins, 59
Comings, 68
Conklin, 32, 34, 36, 39, 40, 41
Cosh, 74
Croisille, 64, 72, 97, 108, 122
Crone, 75
Cuminge, 64, 122, 123
Czołowska, 57

Dalcq, 24, 40
D'Ancona, 108
Daniel, 47
Dantschakoff, 78, 105, 110, 120
Danziger, 64; *65*, *125* (McKay *et al.*)
De Beer, 129, 142
Delbos, 57; *104*, *115* (Cambar *et al.*)
De Smet, 72, 93, 107
Desmond, 137, 138
Detlaff, 24
Deuchar, 86
Devillers, 27, 31, 142
Didier, 109, 110
Dixon, 59, 61, 77, 82, 84, 86, 87, 115
Dodes, 28
Domm, 64, 72, 97, 98, 109, 122

175

Index of animal names

References to figures are in bold type.

179

Subject index

Entries dealing with major topics are in italics; references to figures are in bold type.